# U.S. AIR FORCE YF-23A FLIGHT MANUAL

# U.S. AIR FORCE

# YF-23A FLIGHT MANUAL

**GOVERNMENT REPRINTS PRESS**
Washington, D.C.

Protected under the Berne Convention. Published 2001

Printed in The United States of America
Ross & Perry, Inc. Publishers
717 Second St., N.E., Suite 200
Washington, D.C. 20002
Telephone (202) 675-8300
Facsimile (202) 675-8400
info@RossPerry.com

SAN 253-8555

Government Reprints Press Edition 2001

Government Reprints Press is an Imprint of Ross & Perry, Inc.

Library of Congress Control Number: 2001093417

http://www.GPOreprints.com

ISBN 1-931641-64-1

♾ The paper used in this publication meets the requirements for permanence established by the American National Standard for Information Sciences "Permanence of Paper for Printed Library Materials" (ANSI Z39.48-1984).

# NTM 1F-23(Y)A-1

**USAF SERIES**

# YF-23A

**AIRCRAFT**

**AF 87-0800**

**AF 87-0801**

# UTILITY

# FLIGHT MANUAL

PREPARED BY THE NORTHROP CORPORATION

F33657-86-C-2087

<u>SUPERSEDURE NOTICE</u> -- This manual supersedes NTM 1F-23(Y)A-1 dated 1 December 1989. Destroy previous editions.

<u>DISCLOSURE NOTICE</u> -- If this technical manual is approved for release to a foreign government, it is furnished upon the condition that it will not be released to another nation without specific authority of the Department of the Air Force, that it will be used for military purposes only, that individual or corporate rights originating in the information, whether patented or not will be respected, that the recipient will report promptly to the US, any known or suspected compromise, and that the information will be provided substantially the same degree of security afforded it by the Department of Defense of the United States.

<u>WARNING</u> -- This document contains technical data whose export is restricted by the Arms Export Control Act (Title 22, USC, Sec 2751 <u>et seq</u>) or the Export Administration Act of 1979, as amended (Title 50, USC, App 2401 <u>et seq</u>). Violations of these export laws are subject to severe criminal penalties. Disseminate in accordance with provisions of DOD Directive 5230.25.

# 1 JUNE 1990

INSERT LATEST CHANGE PAGES.
DESTROY SUPERSEDED PAGES.

# LIST OF EFFECTIVE PAGES

NOTE: The portion of the text affected by the change is indicated by a vertical line in the outer margins of the page. Change to illustrations are indicated by miniature pointing hands.

Dates of issue for original and changed pages are:

Original . . . 0 . . . 1 Jun 90

## TOTAL NUMBER OF PAGES IN THIS PUBLICATION IS 210 CONSISTING OF THE FOLLOWING:

| Page No. | *Change No. |
|---|---|
| Title . . . . . . . . . . . . | 0 |
| A . . . . . . . . . . . . | 0 |
| Flyleaf 1 - 2 . . . . . . . . | 0 |
| i - iv . . . . . . . . . . . | 0 |
| 1-1 - 1-126 . . . . . . . . | 0 |
| 2-1 - 2-10 . . . . . . . . | 0 |
| 3-1 - 3-42 . . . . . . . . | 0 |
| 5-1 - 5-2 . . . . . . . . . | 0 |
| 6-1 - 6-2 . . . . . . . . . | 0 |
| 7-1 - 7-2 . . . . . . . . . | 0 |
| A-1 - A-6 . . . . . . . . . | 0 |
| Gloss-1 - Gloss-6 . . . . . . . | 0 |
| Index-1 - Index-6 . . . . . . . | 0 |

CURRENT FLIGHT CREW CHECKLIST

NTM 1F-23YA-1CL-1

1 JUNE 1990

# SAFETY AND OPERATIONAL SUPPLEMENT STATUS

This supplement status page is based on information available on the date of this publication. The information may not be current and must be updated by any subsequently dated supplement status pages.

## SUPPLEMENTS IN THIS CHANGE

| Number | Date | Short Title | Section Affected |
|---|---|---|---|
| | | | |

## OUTSTANDING SUPPLEMENTS

| Number | Date | Short Title | Section Affected |
|---|---|---|---|
| | | | |

# TABLE OF CONTENTS

A29-TOC

## SCOPE

This manual contains the necessary information for safe and efficient operation of the YF-23A flight test prototype aircraft. These instructions provide a general knowledge of the aircraft and specific normal and emergency operating procedures. Your experience is recognized; therefore, basic flight principles and detailed description and operation of systems common to all other aircraft are avoided. This manual provides the best possible operating instructions under most circumstances, but is not a substitute for sound judgement. Multiple emergencies, adverse weather, terrain, etc., may require modification of the procedures.

## USE

This publication is intended for use by an experienced and qualified pilot assigned to the YF-23A Flight Test Program.

## FLIGHT AND OPERATING LIMITATIONS

Flight and operating limitations included herein can be supplemented and/or superseded by Supplemental Flight and Operating Limitations, published under separate cover by Northrop System Safety (Orgn. 3891/89).

## PERMISSIBLE OPERATIONS

Permissible operations will be as scheduled by Northrop Flight Test Operations. The flight manual takes a positive approach and normally states what you can do. Unusual operations or configurations are prohibited unless specifically covered herein. Clearance must be obtained before any questionable operation not specifically permitted in this manual, is attempted.

## HOW TO BE ASSURED OF HAVING LATEST DATA

Check the flight manual cover page, the title block of each safety and operational supplement, and the status page attached to safety and operational supplements. Clear up all discrepancies before flight.

## ARRANGEMENT

The manual is divided into seven fairly independent sections to simplify reading straight through or using as a reference manual.

## SAFETY SUPPLEMENTS

Information involving safety will be promptly forwarded to you in a safety supplement. Urgent information is published in an interim safety supplement and electronically transmitted. Formal supplements are mailed. The supplement title block and formal supplement status page should be checked to determine the effect on the manual and other outstanding supplements.

## OPERATIONAL SUPPLEMENTS

Information involving changes to operating procedures will be forwarded to you in operational supplements. The procedure for handling operational supplements is the same as for formal safety supplements.

## CHECKLISTS

The flight manual contains itemized procedures with necessary amplifications. The checklist contains itemized procedures without the amplification. Primary line items in the flight manual and checklist are identical. If a formal safety or operational supplement affects your checklist, the affected checklist page will be attached to the supplement.

## WARNINGS, CAUTIONS, AND NOTES

Warnings, cautions, and notes are held to an absolute minimum. Everything in this manual could be considered a subject for a note, a caution, or a warning. These definitions apply to warnings, cautions, and notes found throughout the manual:

### WARNING

Operating procedures, techniques, etc., which could result in personal injury or loss of life if not carefully followed.

### CAUTION

Operating procedures, techniques, etc., which could result in damage to equipment if not carefully followed.

### NOTE

An operating procedure, technique, etc., which is considered essential to emphasize.

## USE OF WORDS SHALL, WILL, SHOULD, AND MAY

The following definitions apply to use of the words shall, will, should, and may:

Will is used to express a declaration of purpose or simple futurity.

Shall is used to indicate a mandatory requirement.

Should is used to indicate a preferred but not mandatory method of accomplishment.

May is used to indicate an acceptable or suggested means of accomplishment.

## USE OF CODES

An aircraft code is assigned to each aircraft. The text, illustration, procedure, or chart is coded for information only applicable to that aircraft.

| CODES | SERIAL NUMBER |
|-------|---------------|
| [1] | 87-0800 |
| [2] | 87-0801 |

## YOUR RESPONSIBILITY -- LET US KNOW

Every effort is made to keep the flight manual current. Review conferences with operating personnel assure inclusion of the latest data in the manual. We cannot correct an error unless we know of its existence. In this regard, it is essential that you do your part. Comments, corrections, and questions regarding this manual or any phase of the Flight Manual program are welcomed. They are an essential part of the development program. Direct all comments, corrections, or queries to: Northrop Corporation, Aircraft Division, Dept 8810/89, One Northrop Avenue, Hawthorne, CA., 90250-3277.

## PUBLICATION DATE

Currency of information contained in this manual is as of the publication date appearing on the title page.

# TIME COMPLIANCE TECHNICAL ORDERS

The following TCTO are applicable to this flight manual.

| T.O. NUMBER | TITLE | PRODUCTION EFFECTIVITY | RETROFIT EFFECTIVITY |
|-------------|-------|------------------------|----------------------|
|             |       |                        |                      |

A29-SEC I

SECTION I

DESCRIPTION
AND OPERATION

# TABLE OF CONTENTS

## AIRCRAFT

The YF-23A aircraft is a single place, prototype fighter produced by the Northrop/MCAIR team. Sustained supersonic capability is provided by two engines.

For mission planning gross weight, refer to the current Form 365F.

### COCKPIT ARRANGEMENT

The general arrangement of the cockpit is shown in figure 1-1 (Cockpit General Arrangement), figure 1-2 (Instrument Panel), figure 1-3 (Left Console Panel), and figure 1-4 (Right Console Panel).

### WARNINGS, CAUTIONS, AND ADVISORIES

Abbreviations used to explain warnings/cautions advisories, are located at the end of each system description.
• Master caution light (MC),
• Multipurpose color display (MPCD),
• *VOICE* warnings are in italic bold print.
• Warning **Tone** is in bold print.

# COCKPIT GENERAL ARRANGEMENT (TYPICAL)

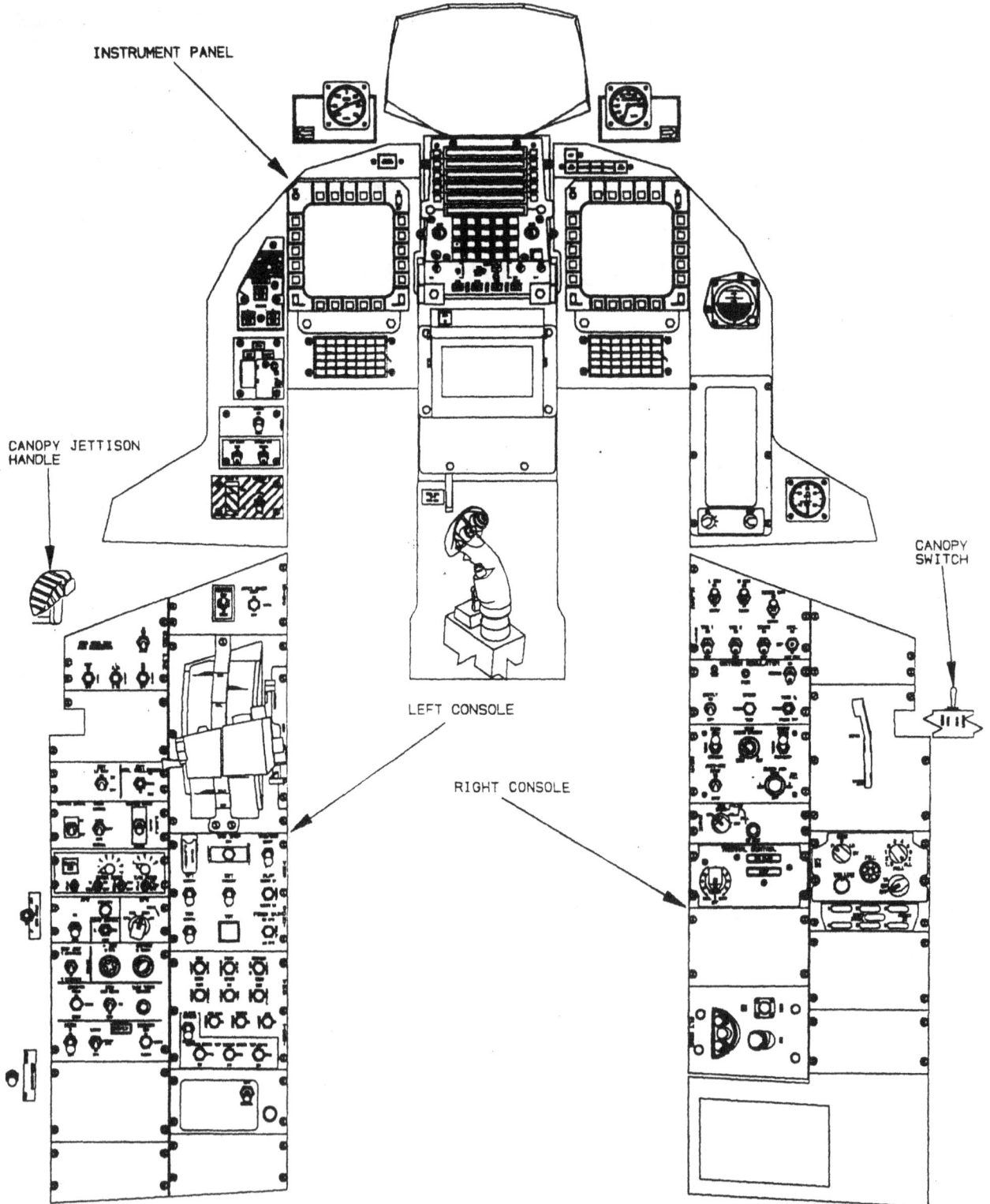

INSTRUMENT PANEL

CANOPY JETTISON
HANDLE

CANOPY
SWITCH

LEFT CONSOLE

RIGHT CONSOLE

A2B-002

*Figure 1-1*

**NORTHROP/MCAIR YF-23A**
**COMPETITION SENSITIVE**
**UNCLASSIFIED**

## *INSTRUMENT PANEL (TYPICAL)*

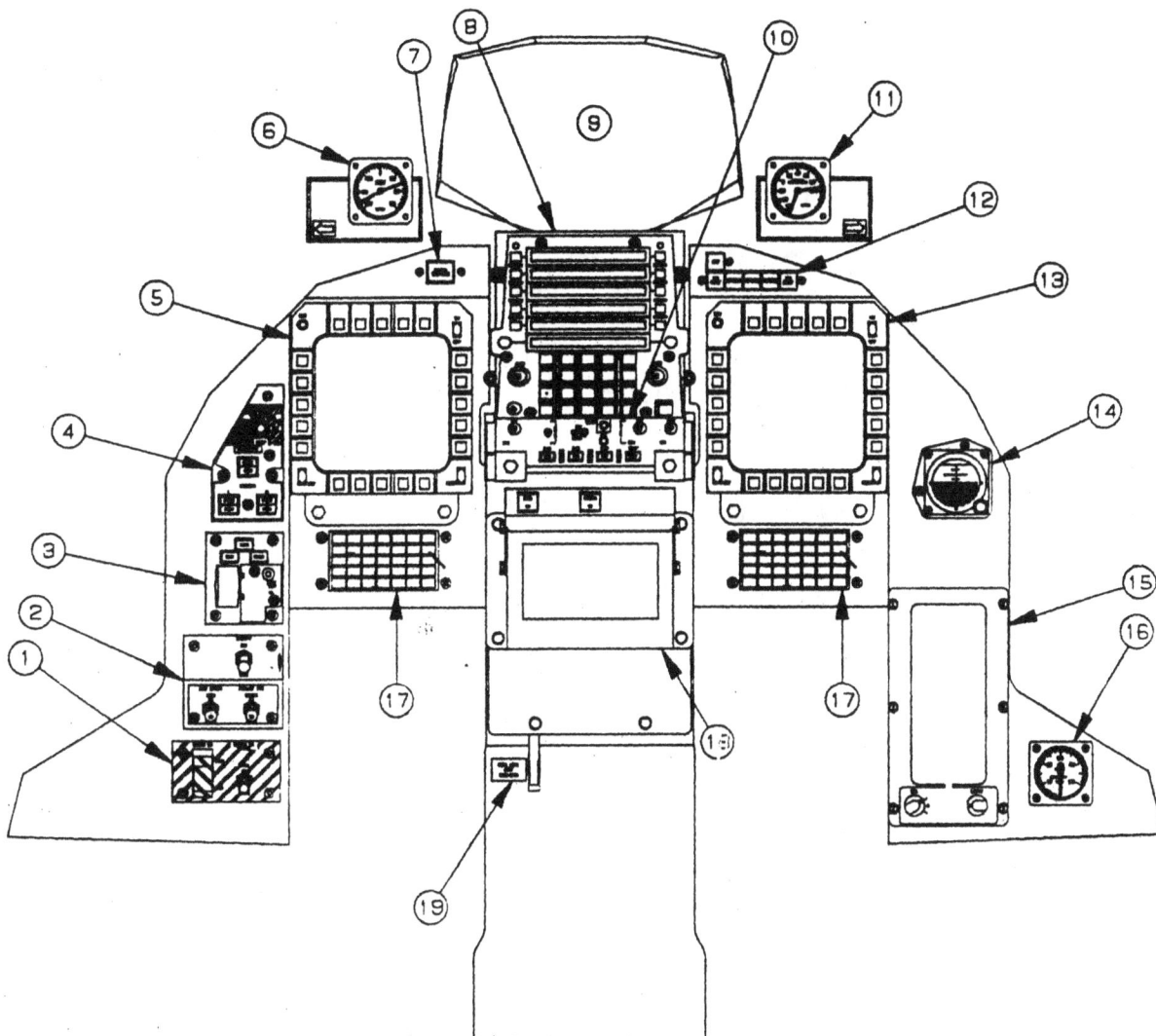

1.  LANDING GEAR AND BRAKES
    EMERGENCY CONTROLS
2.  ARMAMENT CONTROLS
3.  LANDING GEAR CONTROLS
4.  FIRE WARNING AND CONTROLS
5.  MULTIPURPOSE COLOR DISPLAY
    (FLIGHT TEST SPIN PANEL)
    (FLIGHT TEST FLUTTER PANEL)
6.  ANGLE OF SIDE SLIP
    (IF INSTALLED)
7.  MASTER CAUTION LIGHT
8.  UPFRONT CONTROL PANEL
9.  HEAD-UP DISPLAY

10.  HUD CONTROLS
11.  ANGLE OF ATTACK
     (IF INSTALLED)
12.  WARNING LIGHTS
13.  MULTIPURPOSE COLOR DISPLAY
14.  STANDBY ATTITUDE INDICATOR
15.  STANDBY FLIGHT DISPLAY
16.  COCKPIT ALTIMETER
17.  COCKPIT AIR VENT
18.  FLIGHT TEST PANEL
19.  RUDDER PEDAL ADJUST

A29-003

**Figure 1-2**

## LEFT CONSOLE (TYPICAL)

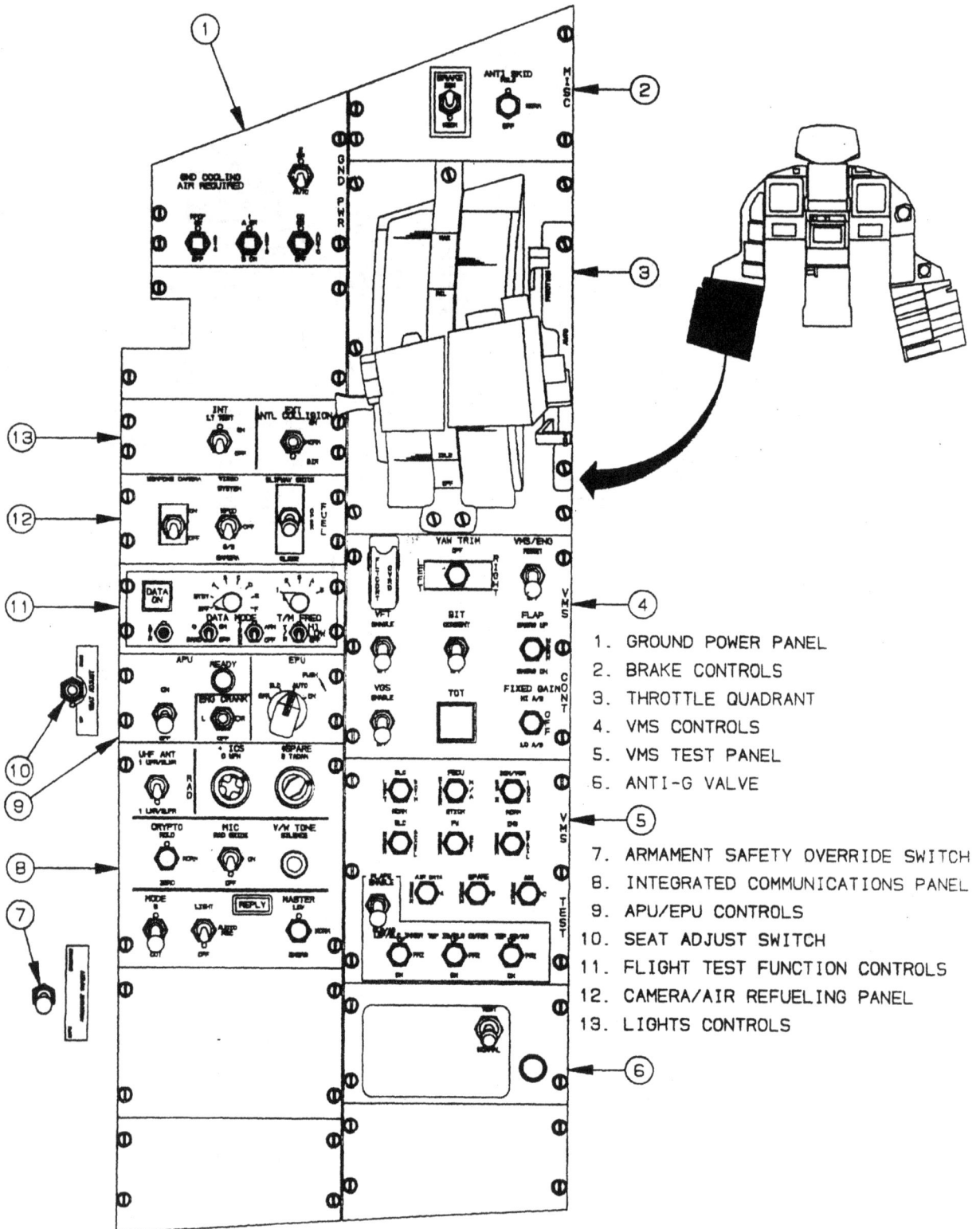

1. GROUND POWER PANEL
2. BRAKE CONTROLS
3. THROTTLE QUADRANT
4. VMS CONTROLS
5. VMS TEST PANEL
6. ANTI-G VALVE
7. ARMAMENT SAFETY OVERRIDE SWITCH
8. INTEGRATED COMMUNICATIONS PANEL
9. APU/EPU CONTROLS
10. SEAT ADJUST SWITCH
11. FLIGHT TEST FUNCTION CONTROLS
12. CAMERA/AIR REFUELING PANEL
13. LIGHTS CONTROLS

*Figure 1-3*

**NORTHROP/MCAIR YF-23A**
**COMPETITION SENSITIVE**
**UNCLASSIFIED**

## *RIGHT CONSOLE (TYPICAL)*

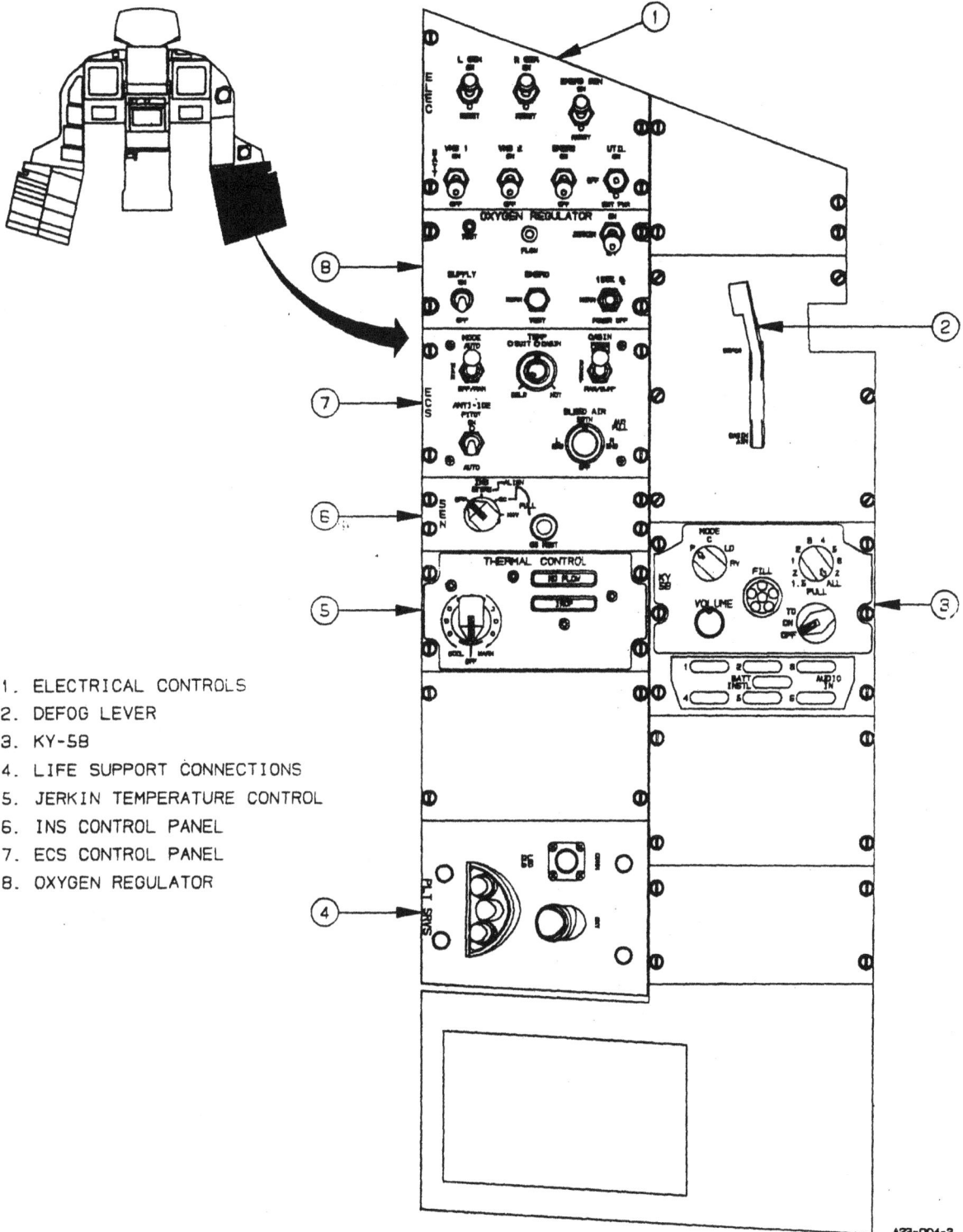

1. ELECTRICAL CONTROLS
2. DEFOG LEVER
3. KY-58
4. LIFE SUPPORT CONNECTIONS
5. JERKIN TEMPERATURE CONTROL
6. INS CONTROL PANEL
7. ECS CONTROL PANEL
8. OXYGEN REGULATOR

A23-004-2

*Figure 1-4*

# [1] PRATT WHITNEY ENGINES

## TABLE OF CONTENTS

## PW ENGINE DESCRIPTION

The aircraft is powered by two twin-spool YF119-PW-100 afterburning turbofan engines.

The engine control system has two Full Authority Digital Electronic Controls (FADEC), with dual electronic sensors and control interface. The engines are controlled electrically by hydraulic valves and actuators. Engine electric and hydraulic systems are self-contained, engine-driven systems. Backup electrical power is available from aircraft power.

The fan and compressor are aerodynamically linked but mechanically independent. Fan speed (N1) and compressor speed (N2) indications are provided.

## ENGINE CONTROL

Computer interface between the aircraft vehicle management system (VMS) computers and the engine mounted FADEC provides engine thrust control. The interchangeable FADEC operate in a dual active mode and have equal authority over control actuators. Degraded engine modes include idle trip, reduced thrust level, partial or total inhibited afterburner, and auto-engine shutdown. The FADEC includes inflight monitoring of engine performance and post-flight maintenance reporting, figure 1-5.

The VMS computers pass Mach and thrust command to the FADEC. The FADEC passes thrust achieved, engine status, and engine performance data to the VMS computers. The VMS computers provide engine performance data via the avionics bus to the engine display.

The FADEC adjusts engine speed and pressure to provide stable, stall-free operation throughout the operating envelope. The FADEC controls:

- Main and afterburner fuel flow.
- Main and afterburner ignition.
- Fan and compressor variable vane angles.
- Exhaust nozzle throat and exit areas.

The FADEC detects engine stall based on compressor discharge pressure. If a stall is detected, engine ignition is activated, fan speed demand is reduced, and pressure ratio demand is reduced. When the stall clears, the engine accelerates to the VMS commanded thrust level.

If engine stall occurs while afterburner is selected, afterburner is terminated. To reselect afterburner after the stall clears, the throttle must be recycled to military power or below.

If communication between the VMS computers and the FADEC is lost for more than two seconds, the engine trips to mid part power (above flight idle).

A switch labeled PW on the VMS test panel permits an increase in military thrust. Allowable N2 increases approximately 2% and EGT 22 degrees Centigrade. Engine operation in this switch position is limited to 5 minutes per flight.

## ENGINE IGNITION

An ignition exciter provides main engine and afterburner ignition in response to FADEC commands. Spark ignition is supplied to the engine by igniter plugs, two in the main combustor and two in the afterburner section. The FADEC commands main engine ignition during engine start, during engine flameout, deceleration, or if engine stall is detected and afterburner ignition when afterburner is selected until lightoff is detected. An engine

## YF-23A/YF119 FADEC SYSTEM OVERVIEW

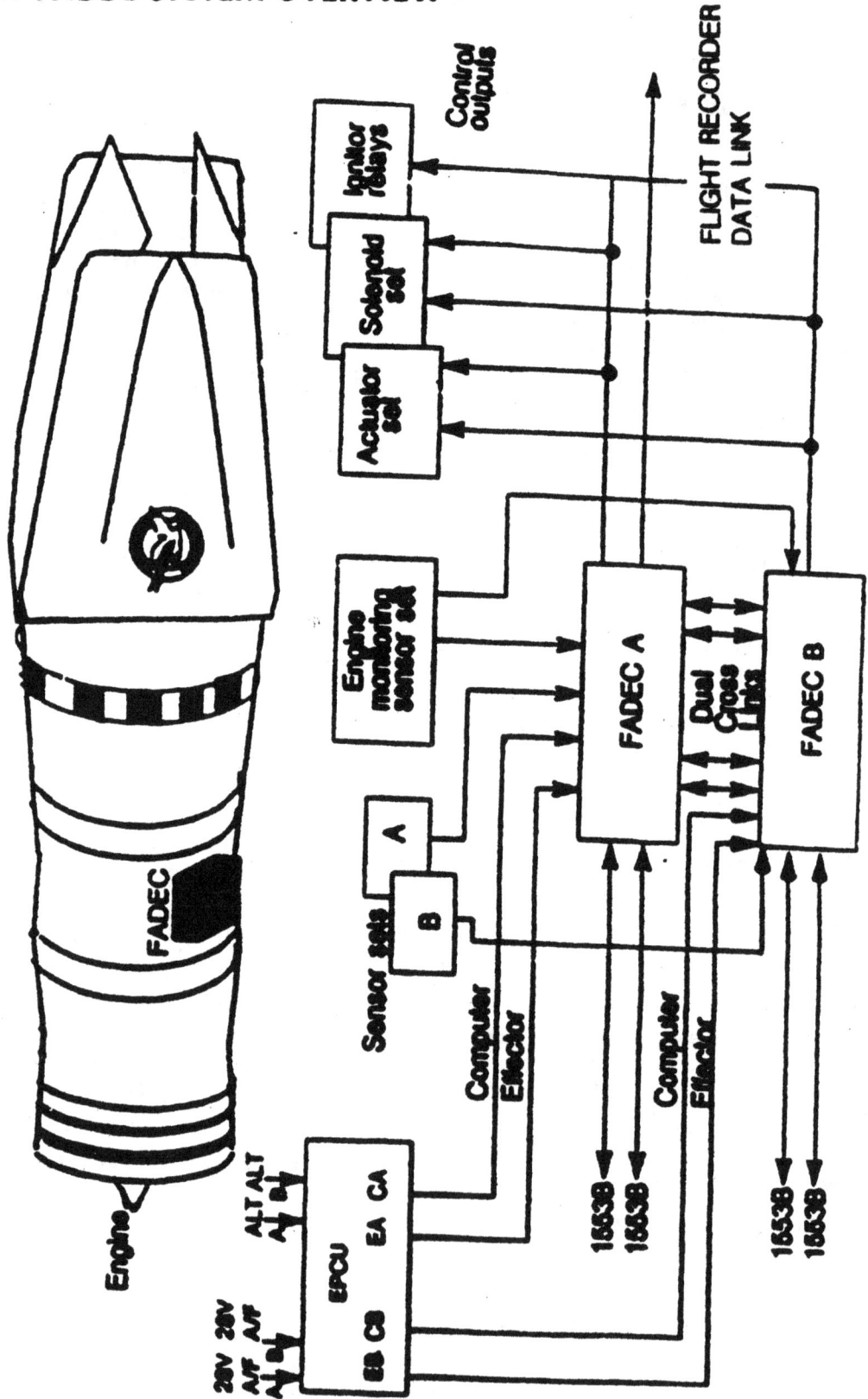

Figure 1-5

mounted alternator provides electrical power for ignition.

## ENGINE FUEL SYSTEM

The engine fuel system (figure 1-6) provides pressurized fuel to the engine combustion section and afterburner. Pressurized fuel is also used for engine hydraulic fluid and lubricating oil cooling and electronic component cooling.

The engine fuel system must have aircraft fuel boost pump pressure to operate properly.

## ENGINE HYDRAULIC SYSTEM

The engine hydraulic system (figure 1-7) uses hydraulic fluid to modulate fuel flow metering and recycling valves, fan and compressor vane actuators, and exhaust nozzle actuators. The system is independent of the aircraft hydraulic system. The major components include the gear box mounted hydraulic pump, reservoir, and actuators for engine and exhaust nozzle control.

## ENGINE OIL SYSTEM

The engine has a self contained, dry sump oil system, figure 1-8. Oil pressure is nonregulated and varies with engine RPM and oil temperature. Oil pressure is indicated at near idle RPM and increases with engine RPM. Oil pressure appears on the standby flight display 1 when dc electrical power is available and on the MPCD engine display when ac electrical power is available.

| CAUTION |

A windmilling engine produces insufficient oil pressure for engine lubrication. Engine damage could occur after 9 minutes.

## ENGINE MOUNTED ACCESSORY DRIVE (EMAD) GEARBOX

The engine mounted accessory drive gearbox drives the engine fuel pump, the oil pressure and scavenge pump, engine hydraulic pump, engine alternator, and power takeoff shaft to the airframe mounted accessory drive unit.

## ENGINE ELECTRICAL SYSTEM

The electrical source for the engine control system, ignition system, and engine electrical accessories is supplied by the engine alternator. The aircraft electrical system provides engine backup power except to the ignition circuits.

## AFTERBURNER

The engine is equipped with a fully modulated afterburner.

Afterburner fuel and ignition are controlled by the FADEC in response to throttle position. Flame sensors in the afterburner section confirm afterburner operation. If the afterburner does not light or a blowout occurs, the FADEC recycles afterburner fuel and ignition three times. Automatic recycling can take from 4 to 20 seconds depending on altitude and number/cause of recycle.

Afterburner failure to light causes a caution ENG 1ST. Cycling the throttle out and back into afterburner range provides afterburner ignition.

A caution ENG 2ND indicates partial or total failure of the afterburner. The affected throttle should be retarded to military power or below and the VMS reset switch placed momentarily to RESET. After reset, afterburner may be used as required.

## VARIABLE EXHAUST NOZZLE

The engine incorporates a two dimensional, convergent-divergent exhaust nozzle. The FADEC controls nozzle throat area for optimum operating pressure within the engine and nozzle exit area for optimum thrust.

## THROTTLE QUADRANT

The throttles (figure 1-9) provide electronic thrust commands through the VMS to the FADEC. The throttles have a detent at military thrust and at minimum afterburner. An idle stop precludes inadvertent engine shutdown. Finger lifts are raised to move the throttles aft of idle.

The throttle mounted switches provide various system controls without moving the hand from the throttle. A throttle friction lever adjusts friction of both throttles. The engine mode lever provides conventional or integrated engine modes.

### Conventional Engine Mode

In the conventional engine mode (CEM), the VMS passes engine thrust commands to the FADEC as a function of throttle position.

### Integrated Engine Mode

The integrated engine mode (ICM) provides airspeed hold. Moving the engine mode lever to AUTO adds a throttle detent at the vertical throttle position labeled CTR and engages VMS integrated engine mode control laws. Putting the throttles in the

# YF119 FUEL CONTROL SYSTEM SCHEMATIC

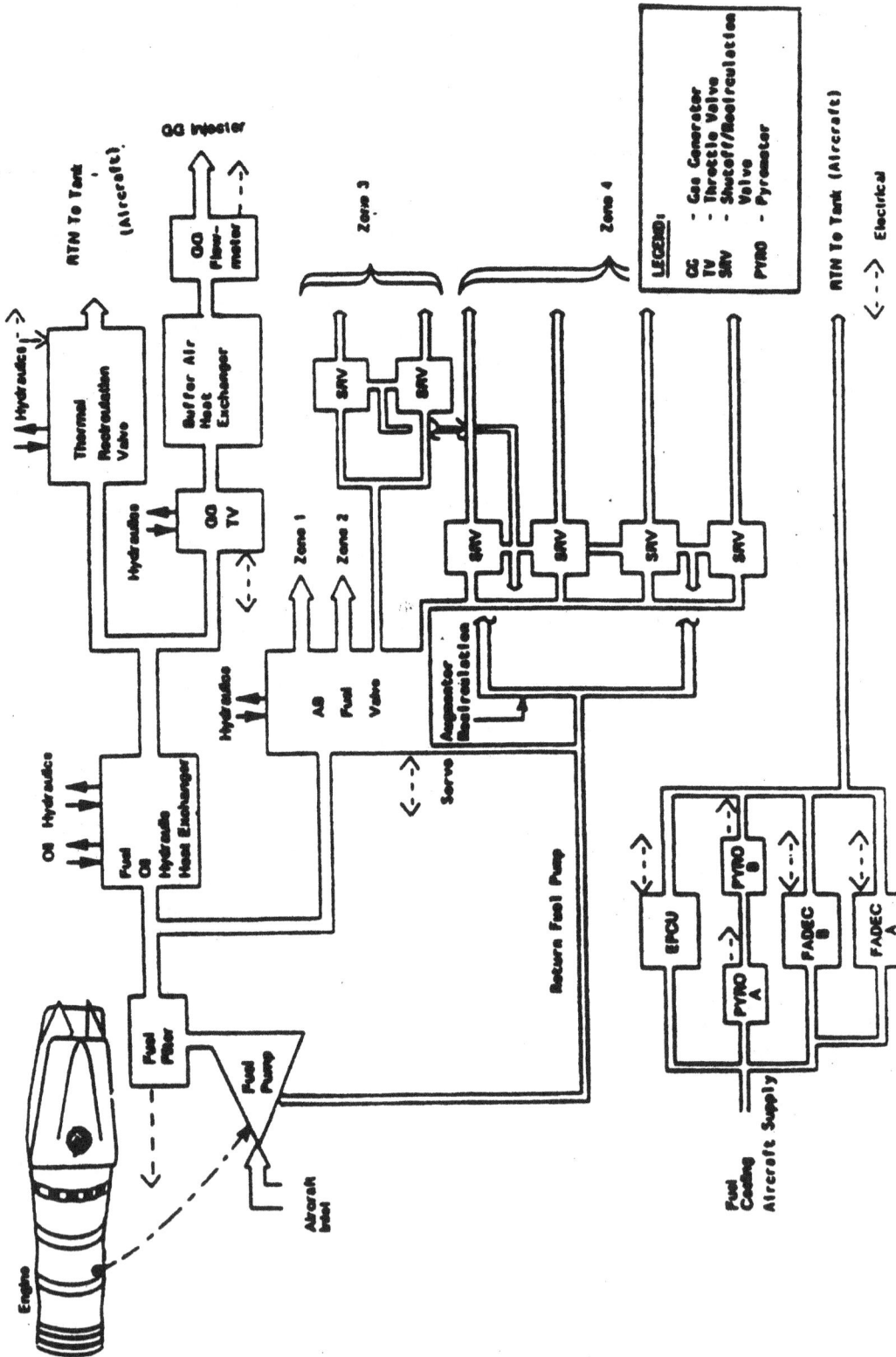

**LEGEND:**

GG : Gas Generator
TV : Throttle Valve
SRV : Shutoff/Recirculation Valve
PYRO - Pyrometer

Figure 1-6

# YF-23A/YF119 HYDRAULIC SYSTEM OVERVIEW

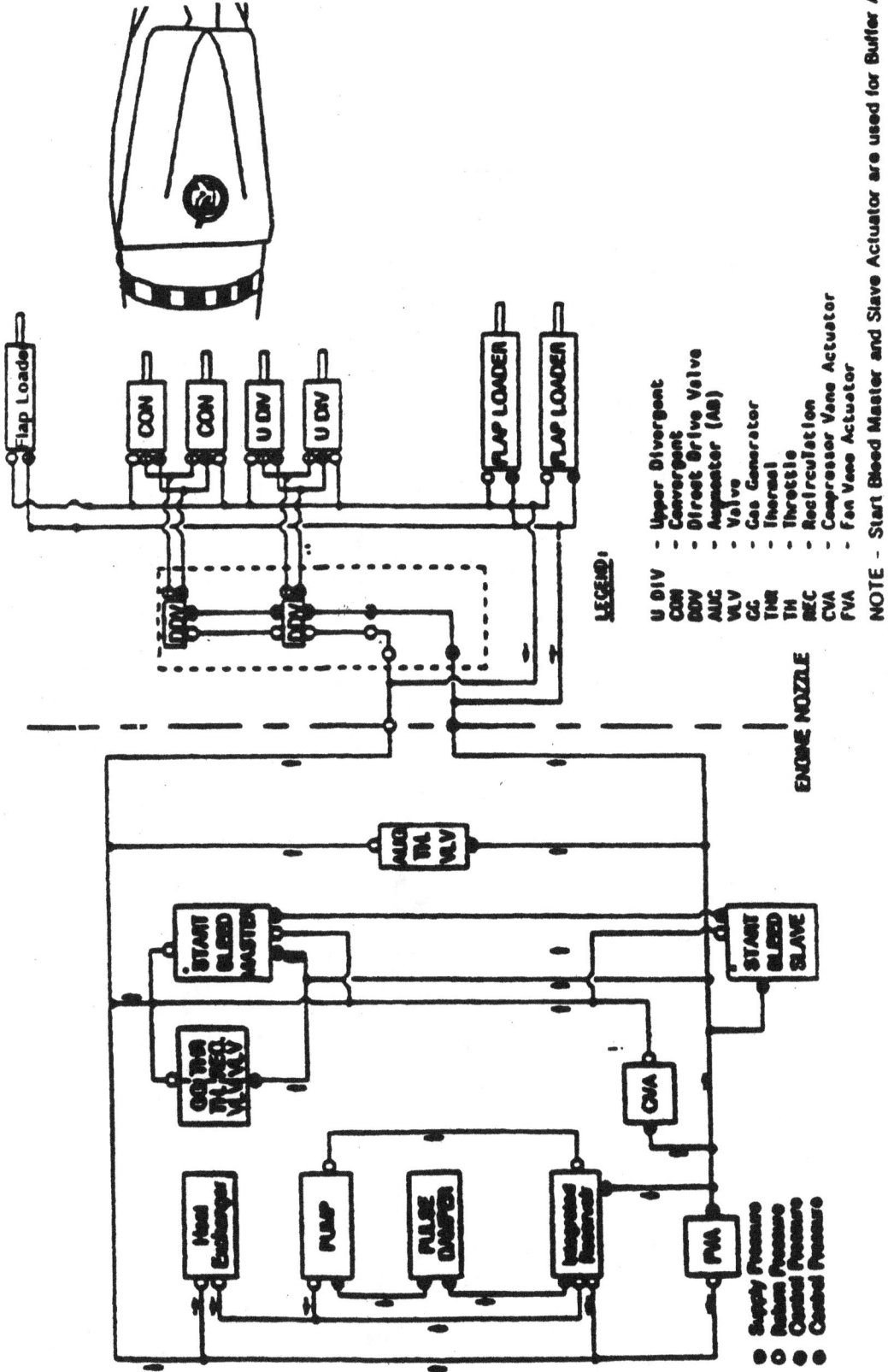

LEGEND:

U DIV - Upper Divergent
CON - Convergent
DDV - Direct Drive Valve
AUG - Augmentor (AB)
VLV - Valve
GG - Gas Generator
TML - Thermal
TH - Throttle
REC - Recirculation
CVA - Compressor Vane Actuator
FVA - Fan Vane Actuator

NOTE - Start Bleed Master and Slave Actuator are used for Buffer Air

Figure 1-7

# YF119 LUBRICATION SYSTEM SCHEMATIC

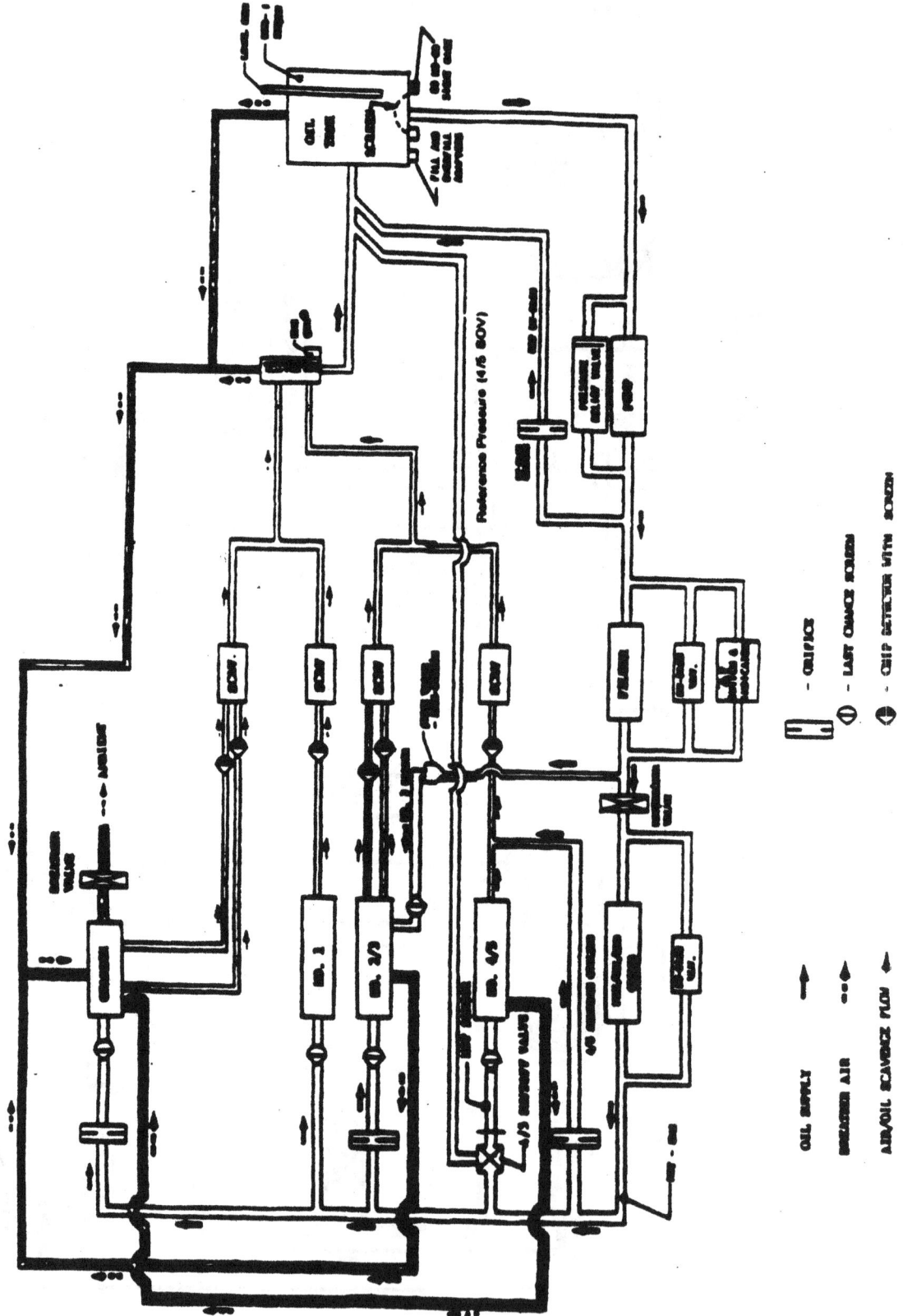

Figure 1-8

## *THROTTLES*

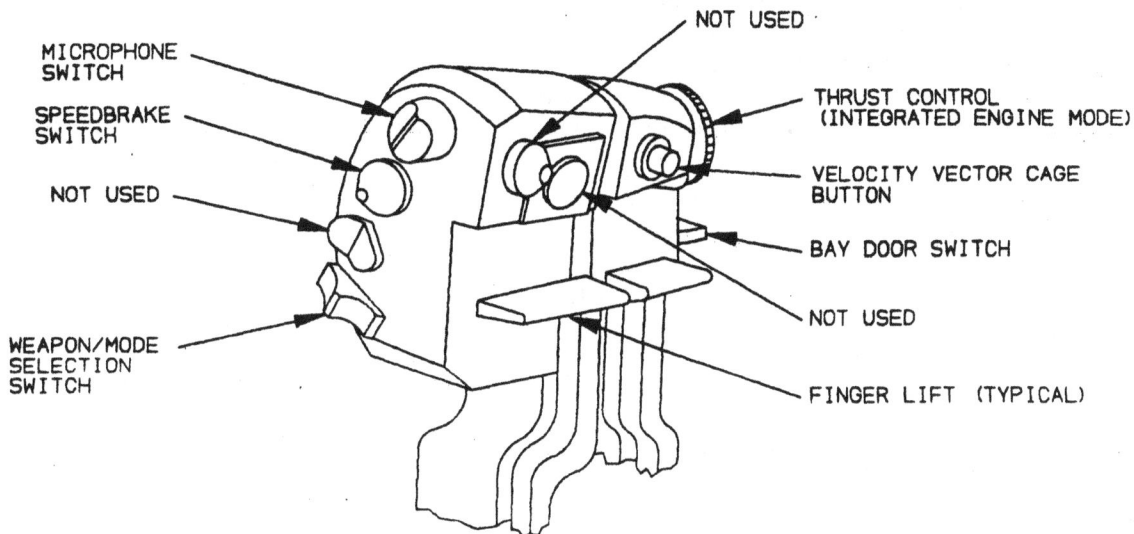

MICROPHONE SWITCH

SPEEDBRAKE SWITCH

NOT USED

WEAPON/MODE SELECTION SWITCH

NOT USED

THRUST CONTROL (INTEGRATED ENGINE MODE)

VELOCITY VECTOR CAGE BUTTON

BAY DOOR SWITCH

NOT USED

FINGER LIFT (TYPICAL)

**THROTTLES**

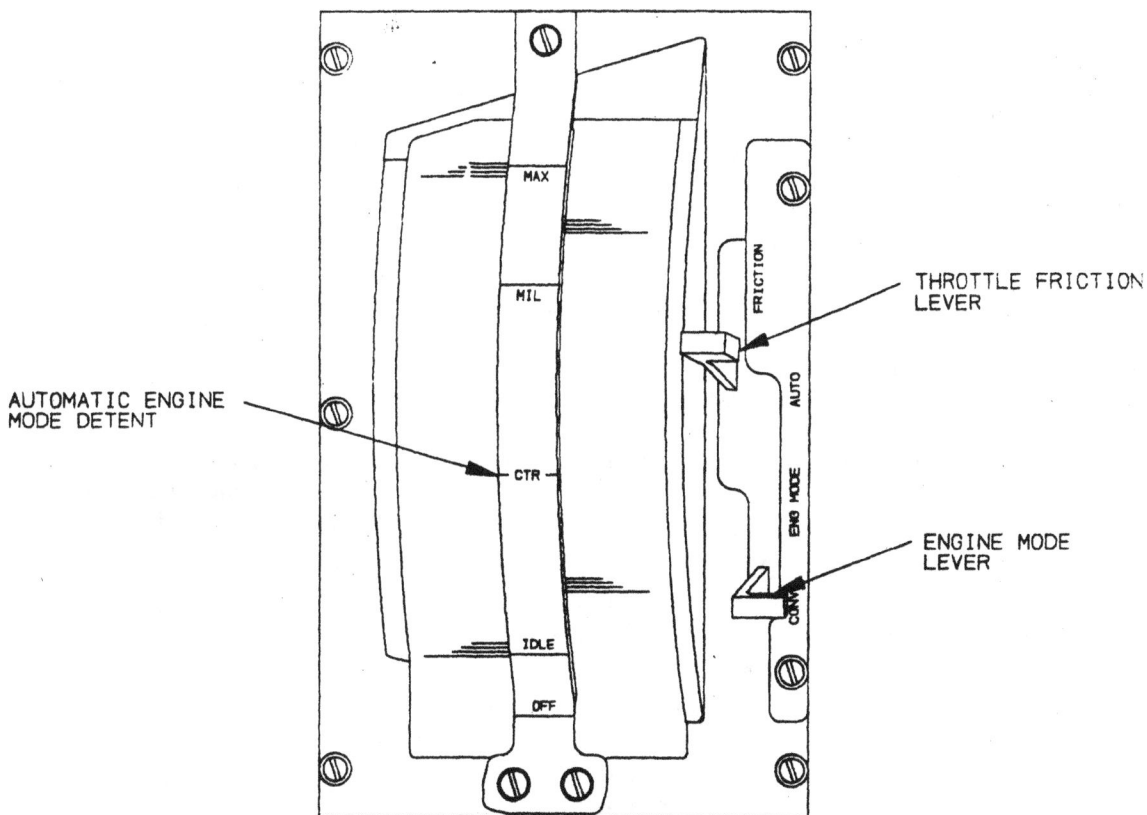

MAX

MIL

CTR

IDLE

OFF

FRICTION

AUTO

ENG MODE

CONV

THROTTLE FRICTION LEVER

ENGINE MODE LEVER

AUTOMATIC ENGINE MODE DETENT

**THROTTLE QUADRANT**
(THROTTLES REMOVED FOR CLARITY)

A23-093

Figure 1-9

detent, holds existing airspeed. The thrust control on the left throttle produces small changes in airspeed. For larger airspeed increases (decreases), one or both throttles can be advanced (retarded) out of the detent. The rate of airspeed change is proportional to the amount the throttle is out of the detent. Both engines respond even if only one throttle is moved. Returning both throttles to the detent holds the new airspeed.

Placing either throttle to idle or military power interrupts the automatic engine mode. Both engines provide thrust based on throttle position as in the conventional mode. Returning both throttles to the detent reengages the integrated engine mode. Either engine in AB reverts engine to conventional engine mode.

## INCREASED STALL MARGIN SWITCH

An increased stall margin switch on the VMS test panel is used to increase engine tolerance to air flow distortion. The switch is label ISM VSM and provides two levels of increased stall margin. Level 1 provides an increase of 50% in stall margin and level 2 provides a 100% increase.

## ENGINE START

The engine crank switch is a three position toggle switch magnetically held in the L or R position (figure 1-10). Placing the engine crank switch to left or right initiates start of the corresponding engine. At the first indication of RPM, the throttle is advanced to idle. The remainder of the start sequence is FADEC controlled. Engine lightoff and EGT rise occur within 20 seconds after 20% RPM. At approximately 53% N2, the starter disengages and the crank switch returns to OFF. Oil pressure appears at 30-60% N2.

Engine start should be aborted for no ignition within 20 seconds after fuel flow, engine RPM hangup below idle, or no oil pressure within 10 seconds after idle RPM. Engine start should be aborted by placing the throttle to OFF and the engine crank switch to OFF.

Engine start should be aborted if maximum EGT exceeded. Prior to idle, if N2 starts to decrease while EGT is still increasing, engine start should be aborted to prevent a hot start. If engine start is aborted for high EGT, the throttle should be placed to OFF while the engine crank switch remains in the L or R position to motor the engine. Motor the engine for 40 seconds.

## ENGINE INDICATIONS

Engine performance indications appear on the standby flight display and the MPCD engine displays, figure 1-11.

### STANDBY FLIGHT DISPLAY 1 AND 2

Standby flight display 1 provides readouts for N2 RPM, EGT, fuel flow, oil pressure, and nozzle position. Indications begin at approximately 8% RPM.

| | |
|---|---|
| RPM HI | N2 in percent from 0 to 109 in 1% increments. |
| EGT | **Exhaust gas temperature in degrees Centigrade from 0 to 1999 in one degree increments.** |
| FF | Fuel flow in hundreds of pounds per hour from 0 to 400 in hundred pound increments. |
| OIL | Oil pressure in pounds per square inch from 0 to 150 in one psi increments. |
| NOZ | Nozzle area ratio in percent. Digital value from 00 to 99 in 1% increments. Pointer from 5 to 95 with marks every 15% increments. |

Standby flight display 2 provides N1 in percent from 0 to 109 in 1% increments.

### MPCD ENGINE DISPLAY

The MPCD engine display is operative when ac power is available. The engine display can be selected on either MPCD with the MPCD button labeled ENG. All legends are white on a black background. If data becomes invalid, the legend OFF replaces the numerical data.

| | |
|---|---|
| RPM % | N2 from 0 to 110% in 1% increments. |
| TEMP °C | Exhaust gas temperature from 100 to 1400 degrees Centigrade in 1 degree increments. |
| FF/PPH | Engine fuel flow in pounds per hour in 10 pound per hour increments. |

## ENGINE CRANK SWITCH

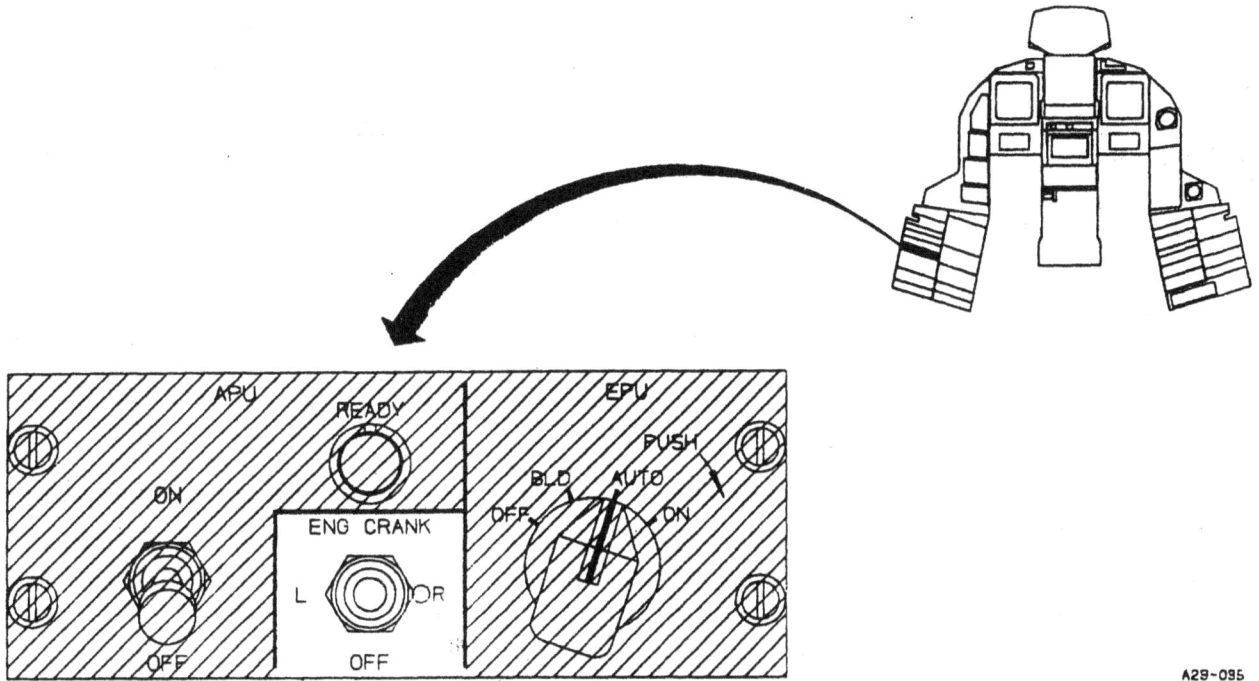

**Figure 1-10**

| | |
|---|---|
| NOZ % | Exhaust nozzle throat area ratio from 0 to 100% open in 1% increments. |
| OIL PSI | Oil pressure from 0 to 100 psi in 1 psi increments. |

## AIRFRAME MOUNTED ACCESSORY DRIVE

An engine power takeoff shaft drives the airframe mounted accessory drive (AMAD). A generator, two hydraulic pumps, and an air turbine starter are mounted on each AMAD.

During engine start, high pressure air from the APU, an operating engine, or an external air source turns the air turbine starter and AMAD. Rotational energy from the AMAD through the power takeoff shaft rotates the engine.

During engine operation rotational energy from the engine drives the AMAD generator and hydraulic pumps.

A self-contained system provides lubrication for the AMAD and cooling for the AMAD, generator, and hydraulic pumps. A fuel/oil heat exchanger cools the

AMAD oil. High AMAD oil temperature causes a caution L(R) AMAD HOT.

## ENGINE CAUTIONS

Engine and AMAD cautions appear on the MPCD and turn on the master caution light. When engine cautions are accompanied by a caution RESET VMS, momentarily placing the VMS/ENG reset switch to RESET, rescinds caution and attempts reset of the malfunction.

| | |
|---|---|
| BYPASS | Boundary layer control door malfunction. |
| L(R) AMAD HOT | AMAD oil temperature hot. |
| L(R) ENG 1ST | • Afterburner failed to relight, |
| **Tone** | • Low oil pressure, |
| | • FADEC and/or cooling fuel overtemperature, |
| | • Loss of control redundancy I, |
| | • Backup power failed. |

## ENGINE INDICATIONS

1. N2
2. EGT
3. FUEL FLOW
4. NOZZLE AREA
5. OIL PRESSURE
6. N1

VALUES SHOWN ARE FOR
ILLUSTRATION PURPOSES
ONLY AND ARE NOT
TYPICAL VALUES

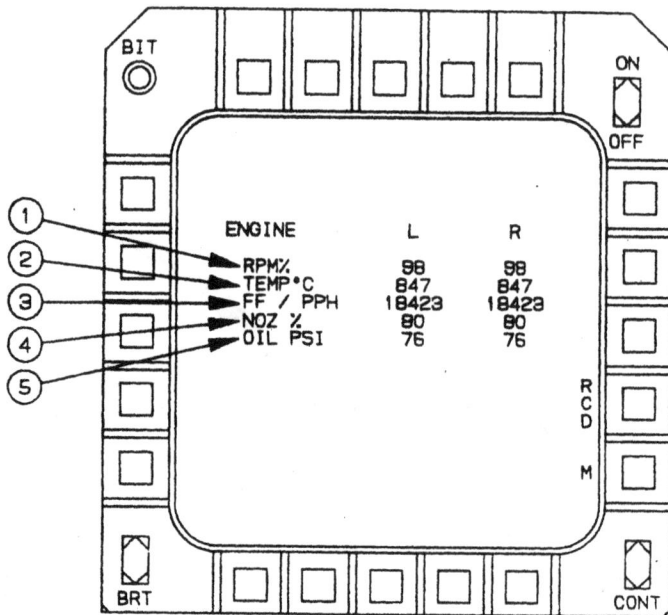

| ENGINE | L | R |
|---|---|---|
| RPM% | 98 | 98 |
| TEMP°C | 847 | 847 |
| FF / PPH | 18423 | 18423 |
| NOZ % | 80 | 80 |
| OIL PSI | 76 | 76 |

MPCD ENGINE DISPLAY

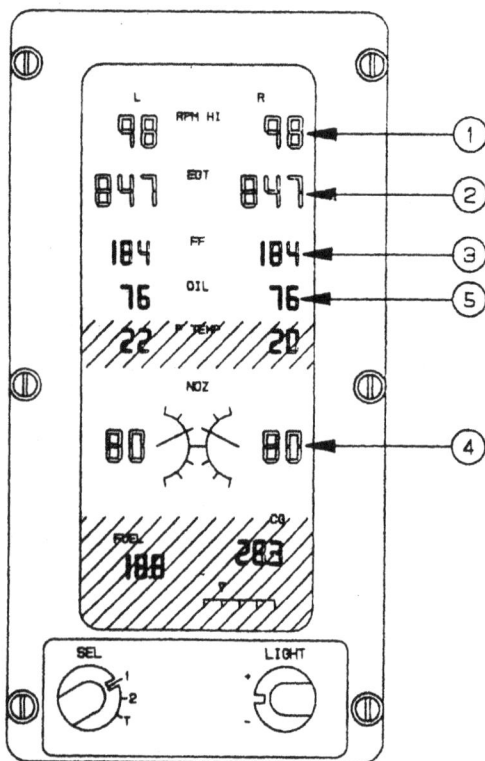

STANDBY FLIGHT DISPLAY 1

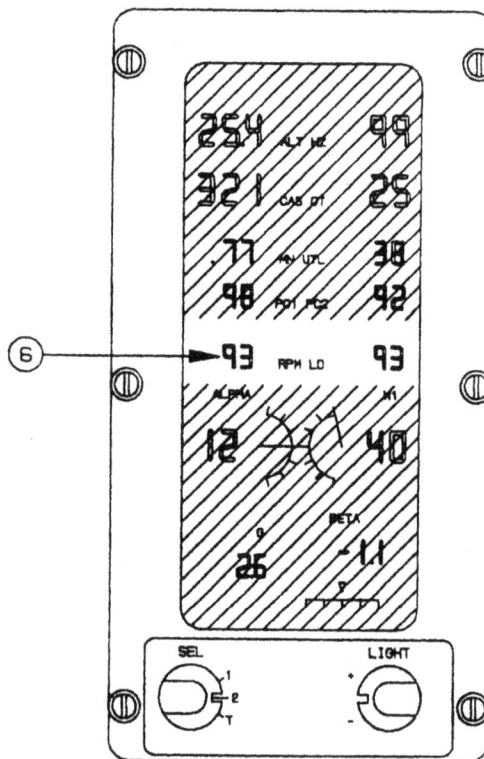

STANDBY FLIGHT DISPLAY 2

A23-084

**Figure 1-11**

**PW/NORTHROP/MCAIR YF-23A**
**COMPETITION SENSITIVE**
**UNCLASSIFIED**

| | |
|---|---|
| **L(R) ENG 2ND** | ● Partial or total afterburner failure, |
| **Tone** | ● Loss of control redundancy II, and FADEC. |
| | |
| **L(R) ENGINE** | ● Zero oil pressure, |
| | ● N1 overspeed, |
| **Tone** | ● N2 overspeed, |
| | ● Engine overtemperature, |
| | ● Uncommanded engine shutdown, |
| | ● Low engine hydraulic pressure, |
| | ● 4-5 buffer air, |
| | ● Engine trip to idle. |

## ENGINE ADVISORY

| | |
|---|---|
| THROTTL | Throttle not operating in selected mode. |

# PW NORMAL PROCEDURES -- ADDITIONAL DATA

Normal Procedures, Section II, engine specific information to Pratt Whitney engines is provided here. Source reference this information is Pratt Whitney integration memo N-90-012, YF-23A/YF119 Engine Flight Manual Addendum, Flight Test Letter #9. The information is provided in the same sequence and under the same headings that are used in Section II.

## EXTERNAL AIR START

If the engine has been run within the previous four (4) hours, motor the engine at 20% RPM for 90 seconds before starting the engine.

## TAKEOFF

If the engines can not be checked individually with throttle at MIL power, advance both throttles to 80% RPM, allow engines to stabilize, and check engine performance prior to brake release.

## LEVEL OFF/CRUISE

Engine performance will vary in flight, depending on outside air temperature, Mach, and altitude. Nozzle position is relatively unaffected by ambient temperatures. Following rapid throttle movement to MIL, RPM could reach 100% RPM before stablizing at a lower operating RPM.

## ENGINE SHUTDOWN/SINGLE ENGINE TAXI

Prior to shutting down an engine, allow engine to idle for five (5) minutes.

# PW EMERGENCY PROCEDURES -- ADDITIONAL DATA

## TABLE OF CONTENTS

Emergency Procedures, Section III, engine specific information to Pratt Whitney engines is provided under this heading. Source reference for this information is Pratt Whitney integration memo N-90-012, YF-23A/YF119 Engine Flight Manual Addendum, Flight Test Letter #9. The information is provided in the same sequence and under the same headings used in Section III.

## AFTERBURNER FAILURE

If an afterburner fails to light or blows out during takeoff (takeoff continued), leave throttle of affected engine in afterburner range. Completion of automatic afterburner recycle is indicated by the caution ENG 1ST if afterburner does not light. The caution ENG 2ND indicates the afterburner use is inhibited.

A low fuel flow and a small nozzle position opening are shown on the MPCD engine display or the standby flight display 1 if the afterburner is not operating.

1. VMS/Engine -- RESET.

2. Afterburner -- Cycle.
   Cycle throttle to MIL and back to afterburner.

## DOUBLE GENERATOR FAILURE -- TAKEOFF

If a double generator failure occurs on takeoff, abort if possible. The potential for a double engine stall exists when the EPU comes on the line and a fuel spike occurs in each engine. If the takeoff must be continued, MIL power provides the best compromise for takeoff/climb power.

If altitude and airspeed permit, the best course of action is to immediately reduce power to 5,000 pounds per engine fuel flow and readvance throttle to MIL after EPU is providing ac power.

1. Throttles -- MIL.

## OUT OF CONTROL RECOVERY

During an out of control condition the throttles should not be moved. If an engine malfunction does exist (stall/flameout/overtemperature), minimize throttle movement and bring engine performance to within parameters.

## ENGINE OIL PRESSURE MALFUNCTION

The caution ENG 1ST is delayed 10-45 seconds for low oil pressure during inverted flight.

If oil pressure is between 5-10 psi, retard throttle affected engine to IDLE and reduce G. If the oil pressure is below 5 psi or above 150 psi, the throttle affected engine should be moved to OFF immediately.

1. Oil Pressure -- Check.
   Check oil pressure on the MPCD engine display or the standby flight display 1.

**If oil pressure is high/low**

2. Throttle -- OFF.

**If oil pressure is 5-10 psi**

2. Throttle -- IDLE.

3. VMS/Engine -- RESET.

## ENGINE STALL

If an engine stall occurs, an automatic engine stall recovery is initiated by the FADEC. Stall recovery action includes afterburner termination, firing main igniters, and rate limited fuel flow. After the stall clears, afterburner must be re-initiated, if required.

A non-recoverable engine stall is indicated by increasing EGT with RPM below 60% and no response to throttle movement. If a non-recoverable stall is indicated, shutdown the affected engine, restart, and leave throttle at idle for the remainder of the flight. If both EGT are high, shut down only one engine at a time to preserve hydraulic system pressure. The engine with the highest EGT should be shutdown and restarted first.

## DOUBLE ENGINE FLAMEOUT

If both engines flameout, immediately check EGT. If EGT is below normal, simultaneous restarts can be attempted.

If the EPU is on the line, airspeed should be 275 KCAS or more for restart. Potential for hot start increases below 275 KCAS.

If EPU is not on the line, maintain 350-375 KCAS to provide 20% minimum RPM. See ENGINE AIRSTART.

## ENGINE AIRSTART

Engine airstart can be accomplished by moving throttle from OFF to IDLE. Moving the throttle OFF, terminates fuel flow and ignition to the engine. Moving the throttle to IDLE activates the FADEC start sequence if engine RPM above 20%.

Optimum engine airstart is 25-50% RPM during engine spooldown.

If RPM drops below 17% RPM, start sequence is terminated. If engine RPM drops to 0%, the engine could seize.

If spooldown conditions can not be met, a windmill or crossbleed airstart can be attempted. Airspeed between 350-375 KCAS provides a minimum 20% engine RPM for start. If a crossbleed airstart is required, maintain 200 KCAS minimum.

## NOZZLE FAILURE

The nozzle position indications are based on the position of the convergent nozzle. If the convergent nozzle fails, the nozzle is hydraulically powered full open and the engine idle trips. VMS/engine reset is required to regain full operation. If the system does not reset, thrust loss at MIL power is minimal.

If the divergent nozzle fails, afterburner operation is inhibited and nozzle position is 58%.

## ENGINE EXHAUST DANGER AREAS

For engine exhaust danger areas, see appendix to interim Section V, Flight and Operating Limitations, published by Northrop System Safety (Orgn 3241/89).

# PW CAUTION ANALYSIS

| INDICATION | CONDITION | CORRECTIVE ACTION |
|---|---|---|
| L ENGINE<br>R ENGINE<br><br>Tone | • Uncommanded engine shutdown.<br>• N1/N2 overspeed.<br>• Engine overtemperature.<br>• Low engine hydraulic pressure.<br>• 4-5 buffer air.<br>• IDLE trip.<br>• Zero oil pressure. | • Check MPCD/standby flight displays.<br>• Reduce power if applicable.<br>• Reset VMS.<br>• Restart engine if applicable.<br>• Shutdown engine if no oil pressure. |
| L ENG 2ND R ENG 2ND<br><br>Tone | • Partial or total afterburner failure.<br>• Loss of control redundancy II.<br>• Degraded capability. | • Check MPCD/standby flight display 1.<br>• Reset VMS. |
| L ENG 1ST R ENG 1ST<br><br>Tone | • Afterburner failed to relight.<br>• Loss of control redundancy I.<br>• Low oil pressure.<br>• FADEC and/or cooling fuel overtemperature. | • Check MPCD/standby flight display 1.<br>• Reset VMS.<br>• Cycle out of afterburner if applicable.<br>• Reduce power and G if low oil pressure. |

# [2] GENERAL ELECTRIC ENGINES

## TABLE OF CONTENTS

## GE ENGINE DESCRIPTION

The aircraft is powered by two twin-spool YF120-GE-100 afterburning turbofan engines.

The fan and compressor are aerodynamically linked but mechanically independent. Fan speed (N1) and compressor speed (N2) indications are provided.

## ENGINE CONTROL

Computer interface between the aircraft vehicle management system (VMS) computers and the engine mounted full authority digital electronic control (FADEC) provides engine thrust control. The triple redundant, engine mounted, FADEC include inflight monitoring of engine performance and post flight maintenance reporting.

The VMS computers pass Mach and thrust command to the FADEC. The FADEC passes thrust achieved, engine status, and engine performance data to the VMS computers. The VMS computers provide engine performance data via the avionics bus to the engine displays.

The FADEC adjusts engine speed and pressure to provide stable, stall-free operation throughout the operating envelope. The FADEC controls:

- Main and afterburner fuel flow.
- Main and afterburner ignition.
- Variable stator vane angle.
- Variable bypass area.

- Exhaust nozzle throat and exit areas.

The FADEC detects engine stall based on compressor discharge pressure. If a stall is detected, engine ignition is activated, fan speed demand is reduced, and pressure ratio demand is reduced. When the stall clears, the engine accelerates to the VMS commanded thrust level.

If communication between the VMS computers and the FADEC is lost, the engine defaults to the lower of last commanded thrust or military thrust.

## ENGINE IGNITION

An ignition exciter provides main engine and afterburner ignition in response to FADEC commands. An engine mounted alternator provides electrical power for ignition.

During engine start, main engine ignition is energized when N2 is above 10% with the throttle out of cutoff. Ignition remains energized until 59% N2.

## ENGINE FUEL SYSTEM

The engine fuel system provides pressurized fuel to the engine combustion section and afterburner. Pressurized fuel is also used for hydraulic and lubrication oil cooling and electronic component cooling.

The engine fuel system must have aircraft fuel boost pump pressure to operate properly.

> **CAUTION**
>
> Avoid use of afterburner with a caution FUEL LOW when below 15,000 feet MSL to prevent fuel pump cavitation and/or engine flameout.

## ENGINE OIL SYSTEM

The engine has a self contained, dry sump oil system. Oil pressure appears on standby flight display 1 with dc electrical power available and on the multipurpose color display (MPCD) engine display with ac electrical power available.

## AFTERBURNER

The engine is equipped with a fully modulated afterburner. Afterburner fuel and ignition are controlled by the FADEC in response to throttle position. If flame sensors in the afterburner section of the engine detect afterburner blowout, the FADEC commands afterburner ignition and minimum afterburner thrust. When the afterburner relights, the thrust advances to the VMS commanded level.

## VARIABLE EXHAUST NOZZLE

The engine has a two dimensional, convergent-divergent exhaust nozzle. The FADEC controls nozzle throat area for optimum operating pressure within the engine and nozzle exit area for optimum thrust.

## THROTTLE QUADRANT

The throttles (figure 1-12) provide electronic thrust commands through the VMS to the FADEC. The throttles have a detent at military thrust and at minimum afterburner. An idle stop precludes inadvertent engine shutdown. Finger lifts are raised to move the throttles aft of idle to the off position.

The throttle mounted switches provide various system controls without moving the hand from the throttle. A throttle friction lever adjusts friction of both throttles. The engine mode lever provides conventional or automatic engine modes.

### CONVENTIONAL ENGINE MODE

In the conventional engine mode, the VMS passes engine thrust commands to the engines as a function of throttle position.

### INTEGRATED ENGINE MODE

The integrated engine mode provides airspeed hold. Moving the engine mode lever to AUTO adds a throttle detent at the vertical throttle position labeled CTR and engages VMS integrated engine mode control laws. Putting the throttles in the detent, holds existing airspeed. The thrust control on the left throttle produces small changes in airspeed. For larger airspeed increases (decreases), one or both throttles can be advanced (retarded) out of the detent. The rate of airspeed change is proportional to the amount the throttle(s) are out of the detent. Both engines respond even if only one throttle is moved. Returning both throttles to the detent holds the new airspeed.

Placing either throttle to idle or military power interrupts the automatic engine mode. Both engines provide thrust based on throttle position as in the conventional mode. Returning both throttles to the detent reengages the integrated engine mode.

## ENGINE START

The engine crank switch (figure 1-13) is a three position toggle switch magnetically held in the L or R position. Placing the engine crank switch to left or right initiates start of the corresponding engine. At 20% N2, the throttle is advanced to idle. The remainder of the start sequence is FADEC controlled. Engine lightoff and EGT rise occur within 10 seconds. Oil pressure appears at approximately 28% N2. At approximately 60% N2 the crank switch is released and returns to OFF.

Engine start should be aborted for no ignition within 20 seconds after fuel flow, N2 hangup below idle, or no oil pressure by 30% N2.

## ENGINE INDICATIONS

Engine performance indications appear on the standby flight displays and the MPCD engine display (figure 1-14).

### STANDBY FLIGHT DISPLAYS

Standby flight display 1 provides engine readouts for N2 RPM, EGT, fuel flow, oil pressure, and nozzle position.

| | |
|---|---|
| RPM HI | N2 in percent from 0 to 109 in 1% increments. |
| EGT | Exhaust gas temperature in degrees Centigrade from 0 to 1999 in 1 degree increments. |
| FF | Fuel flow in hundreds of pounds per hour from 0 to 400 in hundred pound increments. |
| OIL | Oil pressure in pounds per square inch from 0 to 150 in 1 psi increments. |
| NOZ | Nozzle area ratio in percent. Digital value from 00 to 99 in 1% increments. Pointer from 5 to 95 with marks for every 15%. |

Standby flight display 2 provides N1 in percent from 0 to 109 in 1% increments.

### MPCD ENGINE DISPLAY

The MPCD engine display is operative when ac power is available. The engine display can be selected on either MPCD with the MPCD button

## THROTTLES

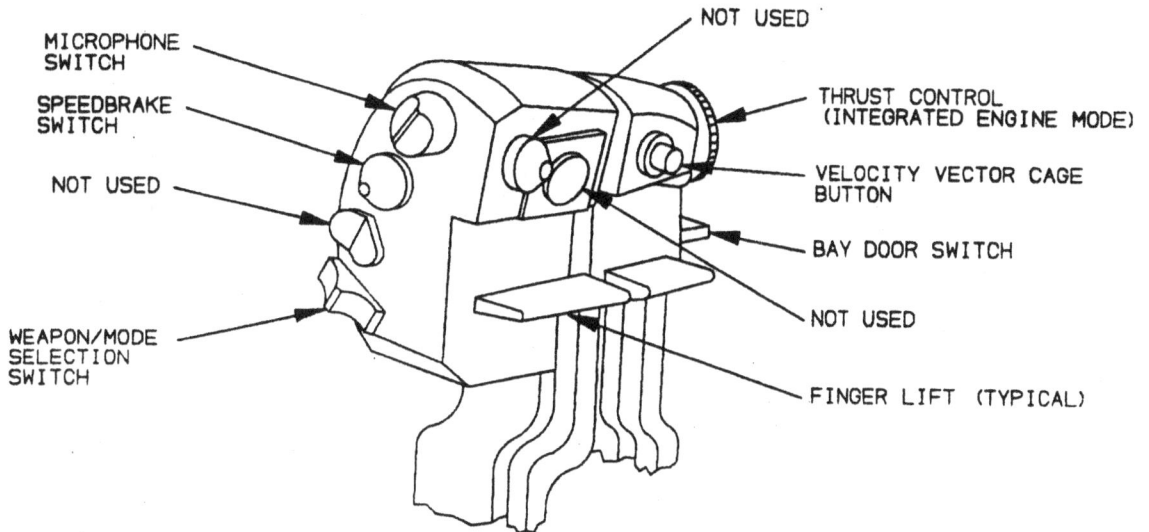

MICROPHONE
SWITCH

SPEEDBRAKE
SWITCH

NOT USED

WEAPON/MODE
SELECTION
SWITCH

NOT USED

THRUST CONTROL
(INTEGRATED ENGINE MODE)

VELOCITY VECTOR CAGE
BUTTON

BAY DOOR SWITCH

NOT USED

FINGER LIFT (TYPICAL)

THROTTLES

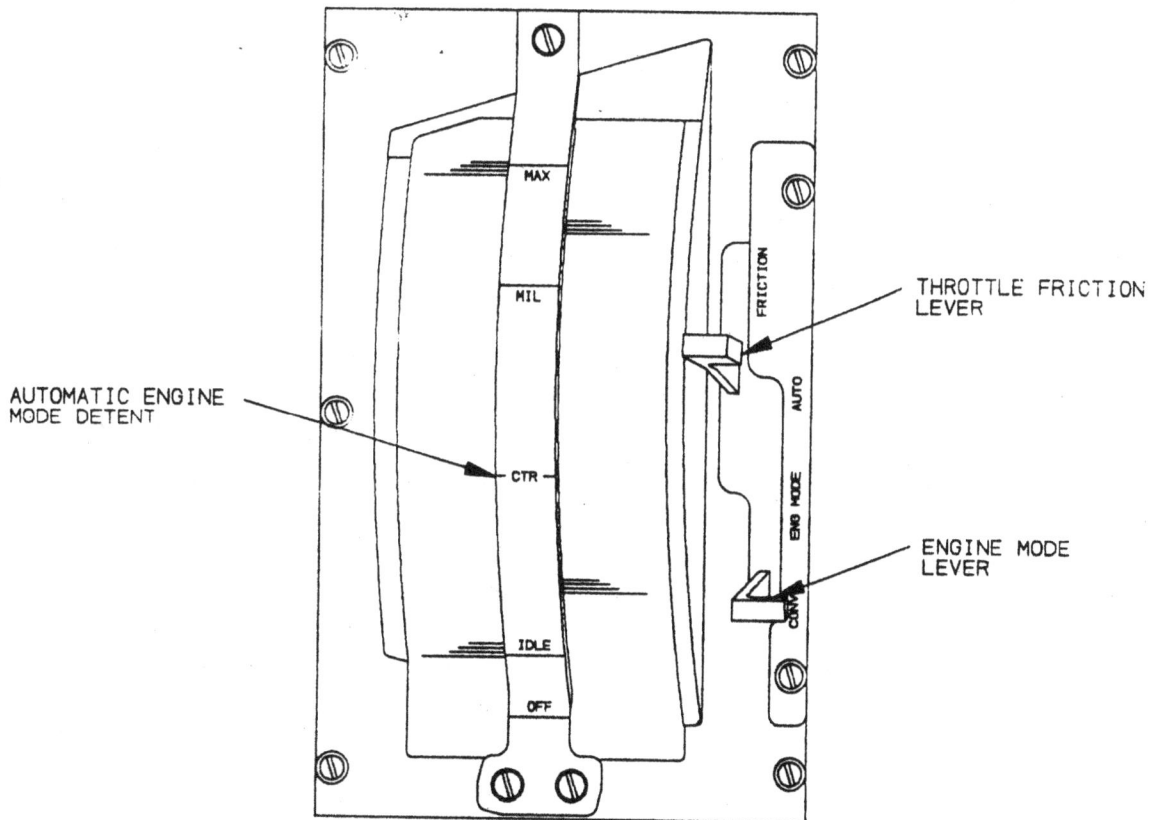

AUTOMATIC ENGINE
MODE DETENT

THROTTLE FRICTION
LEVER

ENGINE MODE
LEVER

MAX

MIL

CTR

IDLE

OFF

FRICTION

AUTO

ENG MODE

CONV

THROTTLE QUADRANT
(THROTTLES REMOVED FOR CLARITY)

A29-099

Figure 1-12

## *ENGINE CRANK SWITCH*

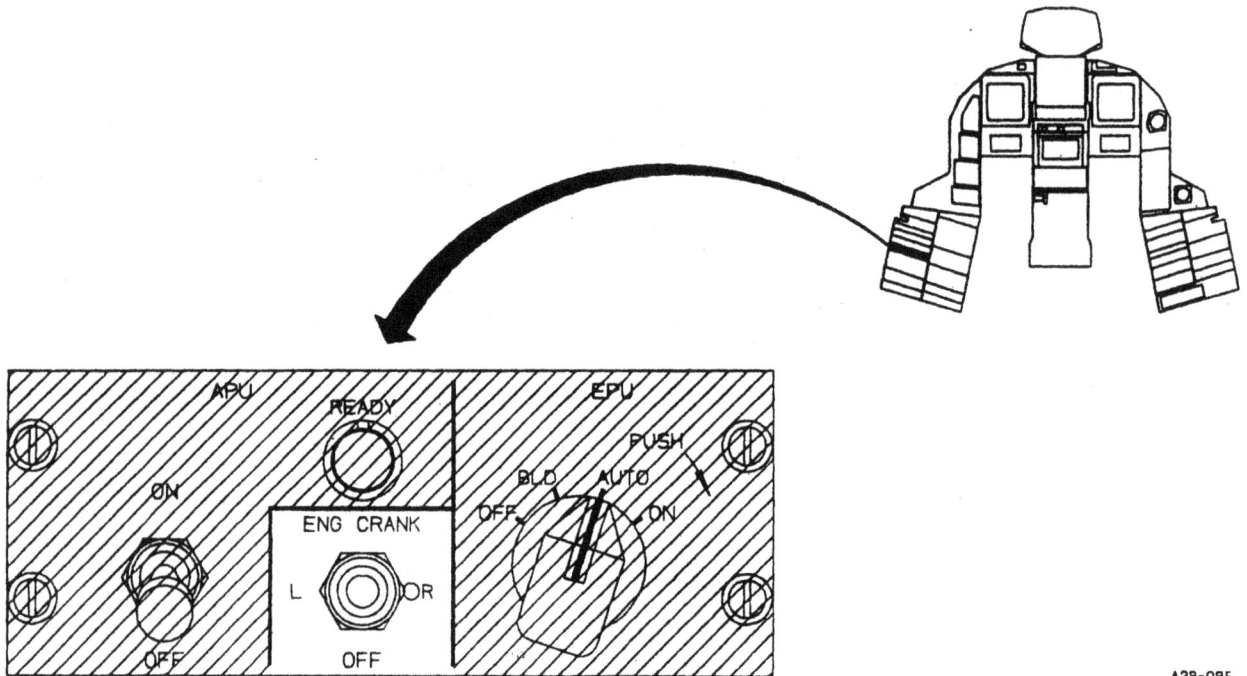

*Figure 1-13*

labeled ENG. All legends are white on a black background. If data becomes invalid, the legend OFF replaces the numerical data.

| | |
|---|---|
| RPM % | N2 from 0 to 110% in 1% increments. |
| TEMP °C | Exhaust gas temperature from 100 to 1400 degrees Centigrade in 1 degree increments. |
| FF/PPH | Engine fuel flow in pounds per hour in 10 pound per hour increments. |
| NOZ % | Exhaust nozzle throat area ratio from 0 to 100% open in 1% increments. |
| OIL PSI | Oil pressure from 0 to 100 psi in 1 psi increments. |

## AIRFRAME MOUNTED ACCESSORY DRIVE

An engine power takeoff shaft drives the airframe mounted accessory drive (AMAD). A generator, two hydraulic pumps, and an air turbine starter are mounted on each AMAD.

During engine start, high pressure air from the APU, an operating engine, or an external air source turns the air turbine starter and AMAD. Rotational energy from the AMAD through the power takeoff shaft rotates the engine.

During engine operation, rotational energy from the engine drives the AMAD generator and hydraulic pumps.

A self-contained system provides lubrication for the AMAD and cooling for the AMAD, generator, and hydraulic pumps. A fuel/oil heat exchanger cools the AMAD oil. High AMAD oil temperature causes a caution L(R) AMAD HOT.

## ENGINE INDICATIONS

1. N2
2. EGT
3. FUEL FLOW
4. NOZZLE AREA
5. OIL PRESSURE
6. N1

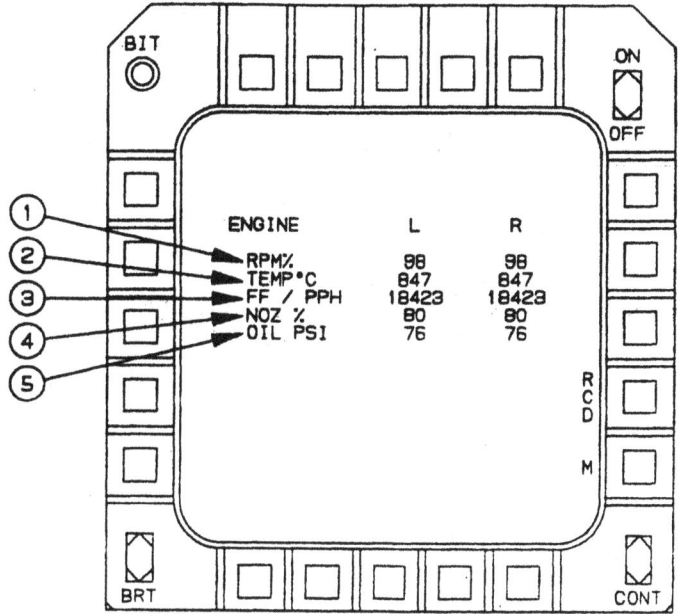

VALUES SHOWN ARE FOR
ILLUSTRATION PURPOSES
ONLY AND ARE NOT
TYPICAL VALUES

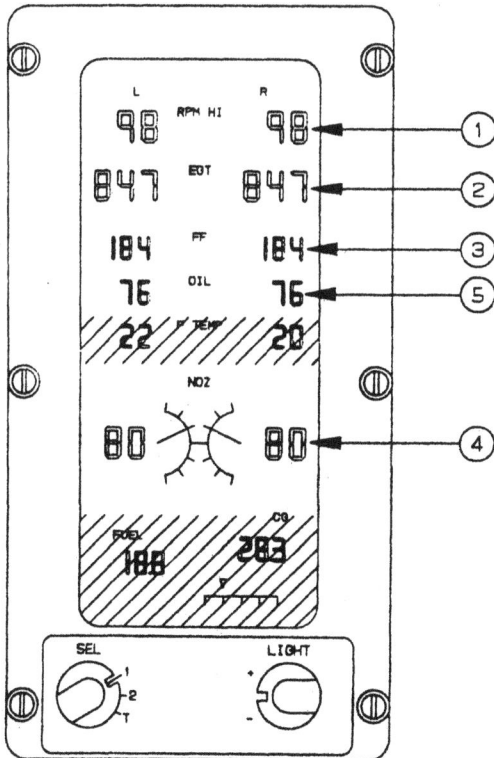

| ENGINE | L | R |
|--------|-----|-----|
| RPM% | 98 | 98 |
| TEMP °C | 847 | 847 |
| FF / PPH | 18423 | 18423 |
| NOZ % | 80 | 80 |
| OIL PSI | 76 | 76 |

MPCD ENGINE DISPLAY

STANDBY FLIGHT DISPLAY 1

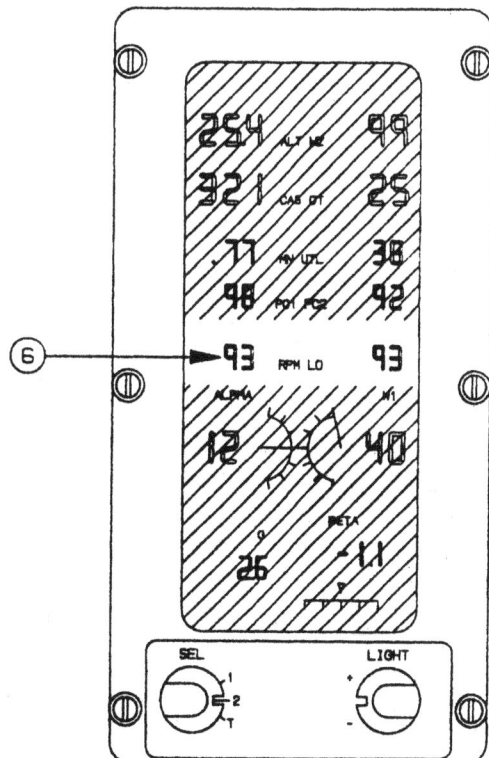

STANDBY FLIGHT DISPLAY 2

A23-034

**Figure 1-14**

## ENGINE CAUTIONS

Engine and AMAD cautions appear on the MPCD and turn on the master caution light. The L(R) ENG 1ST and 2ND cautions are accompanied by a message RESET VMS. Activating the left console VMS/ENG reset switch rescinds the RESET VMS message and attempts reset of the malfunction.

| | |
|---|---|
| BYPASS | Boundary layer control door malfunction. |
| L(R) AMAD HOT | AMAD oil temperature hot. |
| L(R) ENG 1ST | Engine FADEC fault detected. |
| L(R) ENG 2ND | Second like FADEC fault detected. |
| L(R) ENGINE | Engine malfunction or failure. |

## ENGINE ADVISORY

| | |
|---|---|
| THROTTL | Throttle not operating in commanded mode. |

## ENGINE EXHAUST DANGER AREAS

For engine exhaust danger areas, see appendix to interim Section V, Flight and Operating Limitations, published by Northrop System Safety (Orgn 3241/89).

## AUXILIARY POWER UNIT

The gas turbine auxiliary power (APU) unit provides pressurized air for engine starts on the ground. The APU can be used to provide pressurized air to operate the environmental control system on the ground. The APU is located in the APU bay between the engines. The right engine fuel feed system supplies fuel to the APU.

## APU START

The right essential dc electrical bus provides electrical power to start the APU. A hydraulic motor, driven by an APU accumulator, starts the APU. The accumulator is recharged by the utility hydraulic system after engine start.

A spring loaded two position switch, magnetically held to ON, initiates the automatic APU start sequence, figure 1-15. When the APU is ready to accept a load, the green ready light comes on. The APU switch springs back to the position OFF if either throttle is advanced to military power with weight on wheels. The off position stops the APU.

## *APU CONTROL*

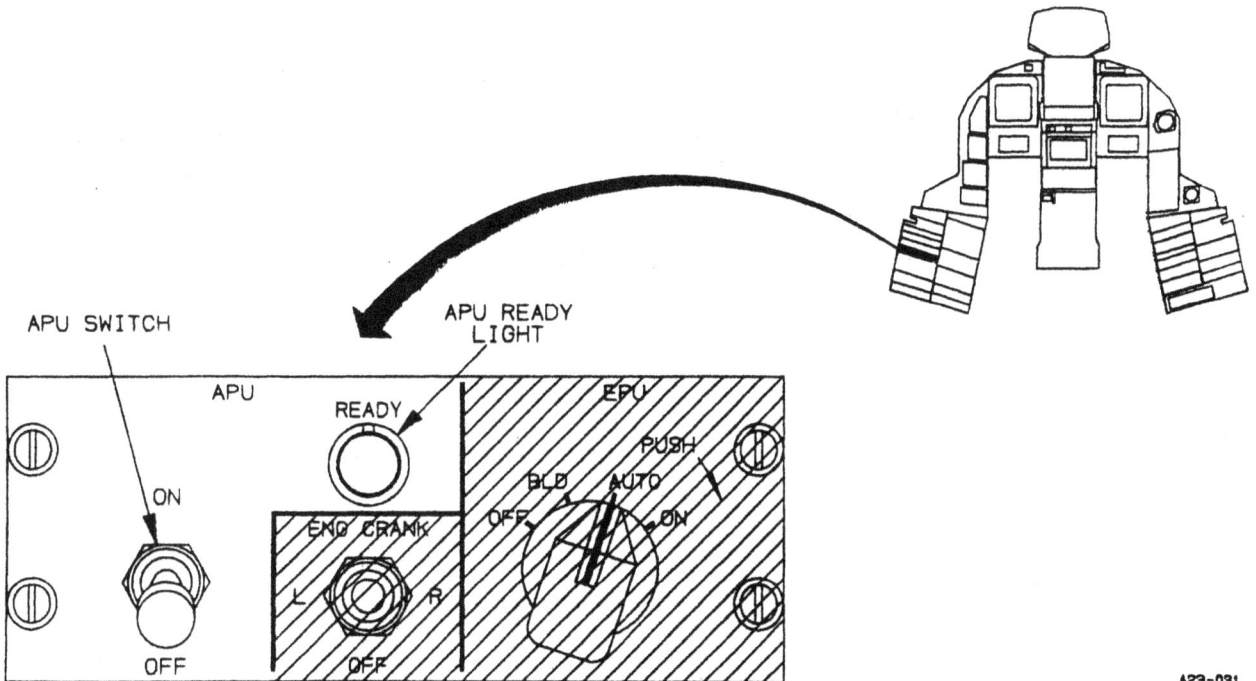

A23-031

**Figure 1-15**

**UNCLASSIFIED**
**NORTHROP/MCAIR YF-23A**
**COMPETITION SENSITIVE**

# ENGINE INLET BOUNDARY LAYER CONTROL SYSTEM

The boundary layer control (BLC) system controls thermal shock and boundary layer interaction on the engine inlet ramp during supersonic flight. Thermal shock and associated air turbulence at the engine face is averted by bleeding off engine inlet ramp boundary layer air. Boundary layer air is bled through a porous section of the inlet ramp and engine ducting and then ducted overboard through two exit doors and a flush mount exit located on the top of the wing. The exit doors are controlled automatically by a schedule stored in the VMS computer.

## ALTERNATE BLC OPERATION

An alternate schedule is provided to improve performance during specific supersonic conditions. The alternate schedule is selected with the BLC accelerate switch, on the VMS test panel. Switch position and test conditions will be provided on flight test card.

## MANUAL BLC OPERATION

Manual operation of the boundary layer control doors is accomplished by selecting switch positions specified in the flight test data card. See VMS Test Panel, this section for switches and description of BLC doors.

# FIRE WARNING AND EXTINGUISHING SYSTEM

The fire warning and extinguishing system provides fire warning and extinguishing for engine, AMAD, or APU bay fires, overheat warning for aircraft structure aft of the engines, and bleed air leak detection and isolation. For bleed air leak detection, see Environmental Control System, this section. The system consists of the fire warning and control panel, two extinguisher bottles, remote sensors, and cockpit overheat/bleed air leak indications. The left and right essential dc busses power the fire warning and extinguishing system.

## FIRE WARNING AND CONTROL

The APU/EPU and the engine (L and R) fire buttons (figure 1-16) have spring loaded metal guards to preclude inadvertent activation. The buttons latch when activated.

### Engine Fire

The top half of an engine fire button indicates a fire in the engine bay. The bottom half indicates a fire in the AMAD bay. An audio *WARNING ENGINE FIRE LEFT* or *WARNING ENGINE FIRE RIGHT* accompanies the corresponding light. The engine fire button shuts off fuel to the engine, shuts off secondary cooling air around the engine, and arms the extinguishing system for discharge to the designated engine or AMAD bay.

### APU/EPU Fire

The APU/EPU fire button indicates a fire in the APU bay. An audio warning *THREAT WARNING* accompanies the light. The button shuts off fuel to the APU, starts a ten second timer, and arms the extinguishing system for discharge to the APU bay after the ten second timer has elapsed. The ten second delay allows the APU to stop prior to extinguisher discharge.

## FIRE EXTINGUISHING OPERATION

Pressing an illuminated engine fire warning button arms the fire extinguisher discharge system. Activation of the extinguisher discharge switch discharges extinguisher to the designated location. For engine bay fires, both extinguisher bottles discharge simultaneously. Only one extinguisher bottle discharges for an APU or AMAD bay fire. The second bottle is discharged by pressing the illuminated fire button a second time to recycle the system to an

# FIRE AND OVERHEAT WARNING AND CONTROLS

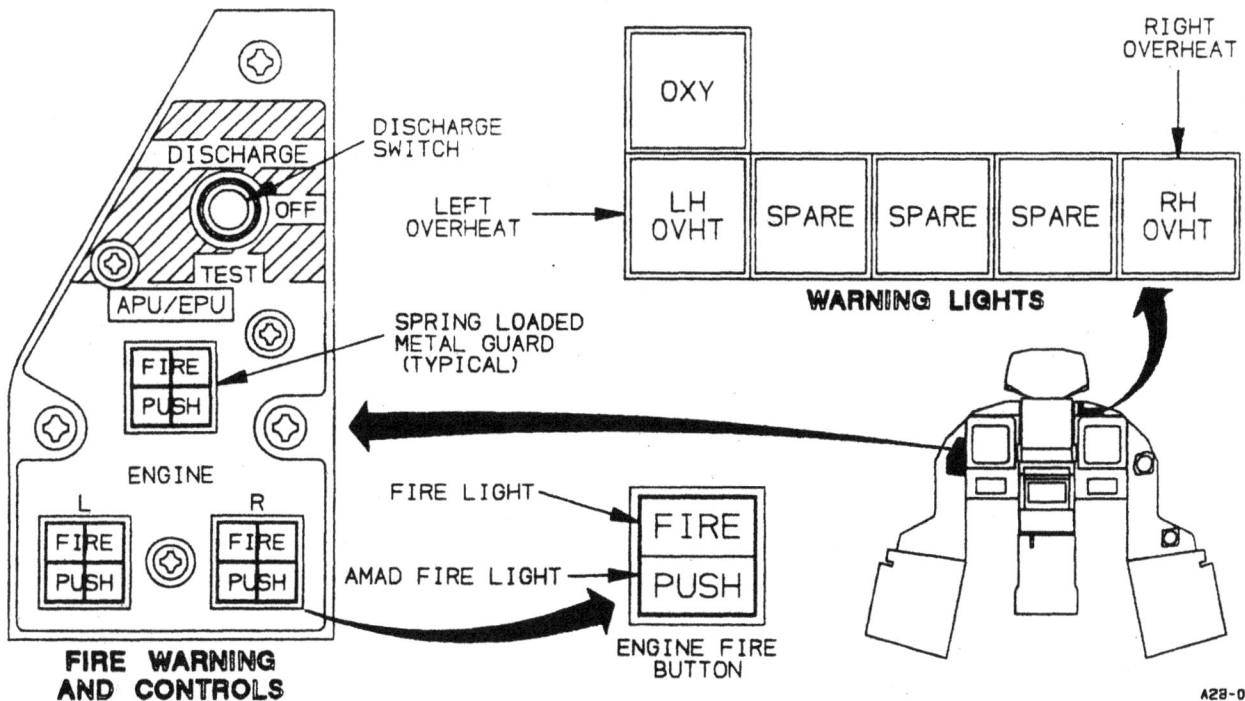

Figure 1-16

armed configuration and then reactivating the extinguisher discharge switch.

## Extinguisher Discharge Switch

The three position extinguisher discharge switch is spring loaded to OFF.

| | |
|---|---|
| DISCHARGE (momentary) | Discharges extinguisher into selected compartment. |
| TEST (momentary) | Tests fire detection, overheat, and bleed air leak sensors. Three fire lights and two over-heat warning lights come on. Bleed air leak warnings appear on the MPCD. |

On the ground, the APU bay has automatic fire protection. The APU/EPU fire light comes on, fuel shuts off, and the extinguisher discharges ten seconds later. If a lighted APU/EPU fire button is pressed, the automatic operation is overridden and the extinguisher discharge switch must be activated to discharge an extinguisher bottle.

## OVERHEAT WARNING

Warning lights labeled LH OVHT or RH OVHT indicate aft fuselage overheat or nozzle burnthrough. An audio warning *WARNING OVERTEMP LEFT* or *WARNING OVERTEMP RIGHT* accompanies the overheat light.

# FUEL SYSTEM

The fuel system for the engines, the APU, and acts as a heat sink cooling hydraulic fluid, generators, AMAD oil, engine oil, and engine components. The fuel management computer (FMC) and transfer system control aircraft CG.

The aircraft has internal fuel tanks: four fuselage and two wing tanks (figure 1-17). Each engine has one feed tank in the forward fuselage body on top of the weapons bay. The center and aft transfer tanks are in the mid-body between the engines. The left and right wings have one transfer tank each.

The fuel system has two ac electrical fuel boost pumps, two dc engine start pumps, four fuel transfer pumps, a crossfeed valve, fuel level sensors, a pressure regulator, pressure relief and vent valves, fuel level control shutoff valves, and an air refueling receptacle. Fuel system information includes fuel caution indications, as well as control and tank quantity indications.

## FUEL TANKS

### Engine Feed Tanks

Each engine feed tank contains an ac boost pump, dc start pump, and the inverted flight compartment. Low fuel sensors are in feed tanks. The right feed tank supplies fuel to the APU.

### Fuselage And Wing Transfer Tanks

The center and aft fuselage tanks are integral and connected by a bulkhead between the tanks. The center tank contains a refueling shutoff valve, vent line, fuel transfer line, and quantity gauging probe. The aft tank contains a small vent tank, and two fuel transfer pumps.

The vent tank captures vented fuel and fuel vapor. An ejector pump returns captured fuel to the aft tank. If the vent tank overfills, excess fuel is dumped overboard through a vent line on the lower aft fuselage, between engines.

The left and right wing tanks are identical and include the smaller duct tanks located over the engine inlet.

All tanks are refueled on the ground through a single point pressure refueling receptacle in the left main gear well. A single point defueling adapter is in the right main gear well.

## FUEL PUMPS

### Fuel Boost Pumps (ac)

Two ac powered boost pumps, one for each engine, supply pressurized fuel to the engines. The boost pumps operate continuously when ac power is available. One pump can supply fuel to both engines through the crossfeed valve.

### Fuel Boost Pumps (dc)

A continuously operating dc fuel pump for each engine supplies pressurized fuel for engine starting (ground and flight), flight idle emergency fuel, and fuel line pressurization with the loss of ac electrical power. The utility battery powers both dc pumps. The dc pump in the right feed tank supplies fuel cooling to the APU and EPU.

### Fuel Transfer Pumps

Four ac powered transfer pumps, two in the aft tank and one in each wing tank, deliver fuel to the engine feed tanks and transfer fuel for CG control.

## FUEL TRANSFER

Fuel transfer is automatic and controlled by the FMC. Transfer pumps in the aft fuselage and wing tanks move fuel into the transfer manifold and forward to the engine feed tanks (figure 1-18). Wing tanks begin transferring first followed by the aft tank. When the aft tank is empty the center tank begins to transfer. After the center tank empties, only feed tank fuel remains.

The FMC cycles the wing tank transfer pumps on and off to balance wing fuel. If the quantity in one wing tank exceeds the other by 300 pounds, the FMC turns the lighter wing transfer pump off to balance wing fuel. During high fuel flows, both wing transfer pumps operate regardless of wing fuel imbalance.

### Gravity Transfer

Wing fuel gravity flows from the wing tanks to the aft tank if the transfer pumps fail. Fuel gravity flows from the aft tank to the center tank through an interconnect valve. Fuel gravity flows from the center tank to the feed tanks through flapper valves.

### FUEL PRESSURIZATION AND VENT SYSTEM

The fuel pressurization and vent system maintains fuel tank pressure to prevent fuel pump cavitation (figure 1-19). The system consists of an air pressure regulator valve, vent relief valves, flame arrestor, dive vent check valve, and air check valves.

## FUEL TANKS

VENT TANK

ENGINE SHUTOFF
VALVE

CELL 6A

CROSSFEED VALVE

ENGINE
BOOST
PUMP

CELL 6B

CELL 4

CELL 2B

ENGINE SHUTOFF
VALVE

CELL 3

CELL 2A

CELL 5B

TRANSFER PUMP
(4 PLACES)

CELL 5A

CELL 1A

CELL 1B ———————————— ENGINE BOOST PUMP

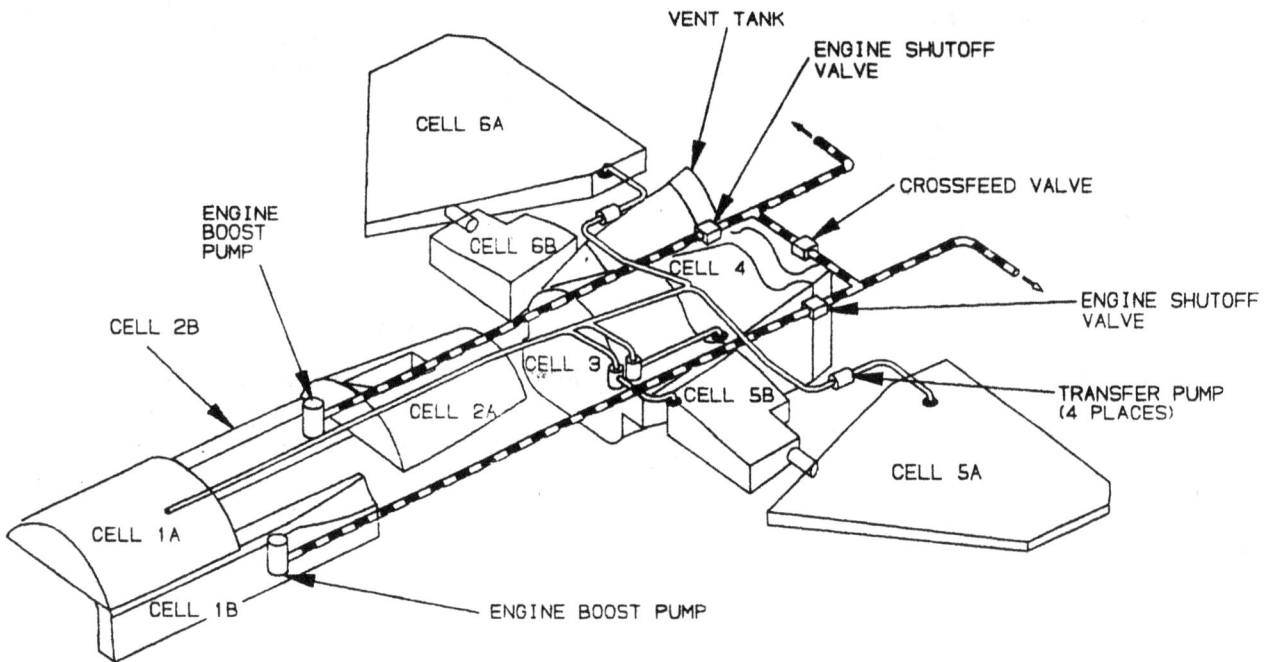

LEFT ENGINE FEED TANK - CELL 1A AND CELL 1B
RIGHT ENGINE FEED TANK - CELL 2A AND CELL 2B
CENTER FUSELAGE TANK - CELL 3
AFT FUSELAGE TANK - CELL 4
LEFT WING TANK - ECLL 5A AND CELL 5B
RIGHT WING TANK - CELL 6A AND CELL 6B

====== FUEL TRANSFER

▆▄▆▄ ENGINE FEED

A23-051

**Figure 1-17**

## FUEL TRANSFER/REFUEL

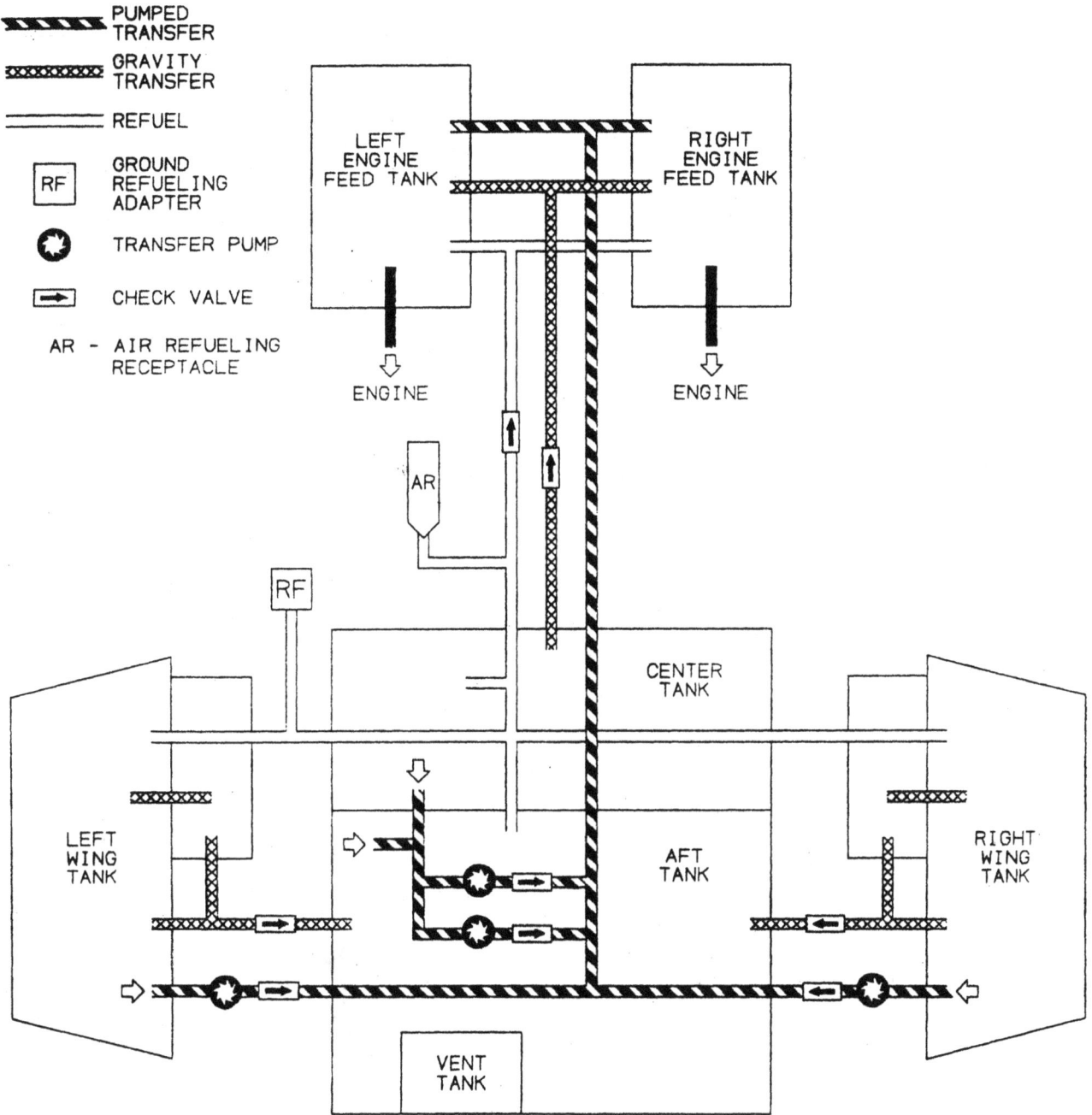

**Figure 1-18**

## FUEL PRESSURIZATION/VENT

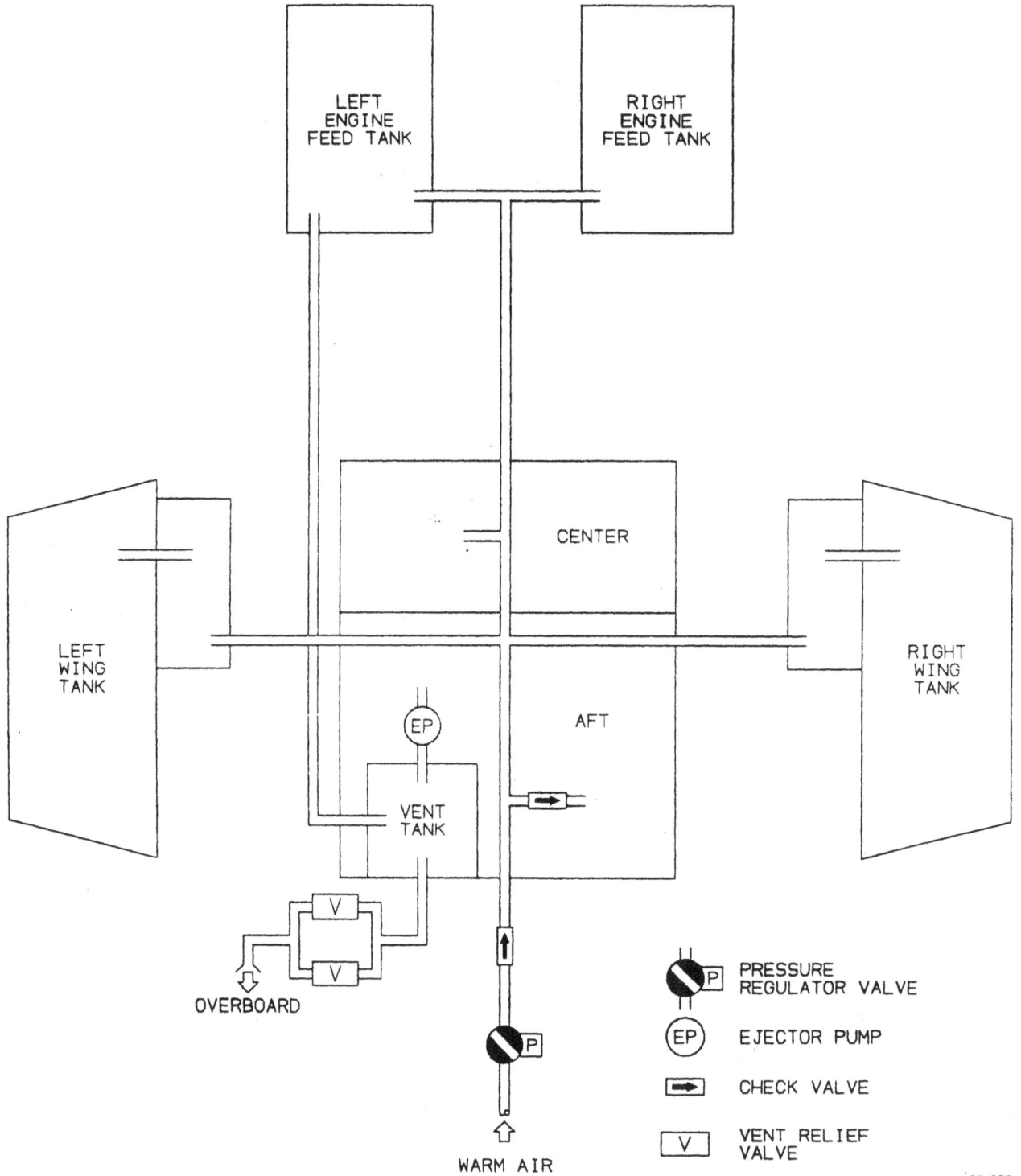

Figure 1-19

Engine bleed air maintains fuel tank pressure above the fuel vapor pressure. The fuel pressure regulator reduces bleed air pressure to maintain fuel tank pressure approximately 0.4 PSI above ambient pressure up to approximately 17,000 feet. Above 17,000 feet, the regulator maintains the tank pressure between 7.5 PSI and 8.3 PSI absolute.

Dual redundant vent tank relief valves provide tank overpressure protection and dive venting (collapsing pressure) protection.

## THERMAL MANAGEMENT SYSTEM

### Fuel Heat Exchange

The fuel system uses heat exchangers to transfer heat to the fuel. Engine feed fuel cools the electronic engine control, then returns to the feed tanks (figure 1-20).

AMAD/equipment oil heat exchangers transfer heat from the hydraulic system oil, generator cooling oil, and AMAD oil to engine feed fuel. After passing through the AMAD/equipment oil heat exchangers, fuel passes through engine oil heat exchangers.

Each engine feed line has a temperature sensor and fuel bypass valve. The FMC monitors engine fuel inlet temperature. If engine fuel inlet temperature reaches 104 degrees C, the FMC opens the bypass valve for thermal fuel management. If engine fuel inlet temperature reaches 115 degrees C, the caution FUEL HOT is initiated. With the bypass valve open, fuel recirculates to two ram air heat exchangers where intake air cools the fuel. After passing through the ram air heat exchanger, cooled fuel is routed to the wing or engine feed tanks.

[2] During APU operations, APU feedline fuel passes through two EPU oil heat exchangers cooling EPU gearbox oil.

## FUEL SYSTEM CONTROLS AND DISPLAYS

The fuel system controls and displays are the MPCD fuel display and standby flight display 1. Menu-driven fuel system controls and displays are available on either MPCD. The standby flight display provides information to backup the MPCD.

### Standby Flight Display 1

The standby flight display 1 shows the left and right engine fuel flow, engine fuel temperature and the total fuel quantity (figure 1-21). The fuel temperature is measured ahead of the engine spray bars. The caution FUEL HOT, might not appear on the MPCD.

### Fuel Display

The MPCD fuel display enables fuel system checkout, monitor, and control. The fuel display shows tank quantity, total fuel quantity, CG and bingo fuel (figure 1-22).

Individual fuel tank quantity is shown. A moving caret indicates the ratio of fuel available to the fuel tank capacity. Fuel quantity presentations are green. Data on the MPCD changes color from green to amber if the data is invalid.

If a fuel gaging probe in a tank fails, the legend EST appears beside the tank quantity indication. The fuel quantity could be inaccurate due to aircraft attitude and the fuel quantity is estimated.

If both fuel gaging probes in a tank fail, the legend INV appears beside the tank quantity. The fuel quantity for the tank is computed with no fuel remaining and the total fuel quantity is invalid. The time since the fuel quantity was updated appears on the lower left of the display.

If the fuel management computer fails, the fuel quantity indications freeze. The legend INVALID appears next to the time since fuel quantity was last updated. The vehicle management system uses the last fuel quantity update for CG computation.

## FUEL SYSTEM OPERATION

Fuel system operation is normally automatic. Manual operation of the fuel system is provided if automatic operation fails.

### Automatic Fuel Operation

The FMC controls the fuel system automatically. The FMC interfaces with the vehicle management computer (VMC) for fuel and engine thermal management. The FMC interfaces with the avionics central computer to enable fuel system manual control, fuel quantity, and fuel system status.

The FMC powers-up with:

- APU start,
- Engine start,
- Utility Battery switch on.

With external ground power applied to the aircraft, the FMC does not power-up until switch 1 on the ground power panel is selected to A ON or B ON.

### Manual Fuel Operation

Manual control of the fuel system is initiated by pressing the button labeled CONT on the fuel display. The button adjacent to each component label

# *ENGINE FEED/THERMAL MANAGEMENT SYSTEM*

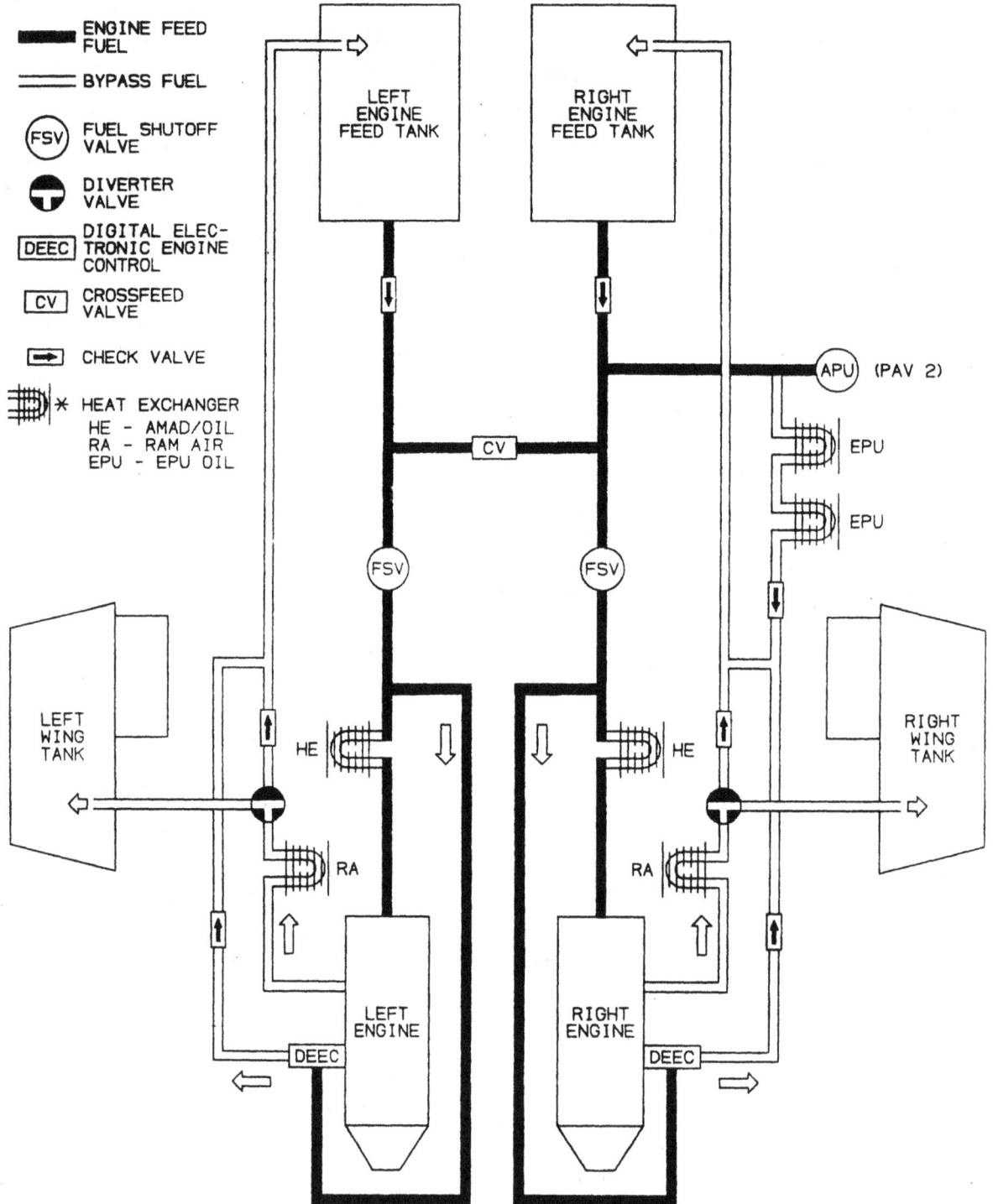

*Figure 1-20*

**NORTHROP/MCAIR YF-23A**
**COMPETITION SENSITIVE**
**UNCLASSIFIED**

## *STANDBY FLIGHT DISPLAY 1 -- FUEL*

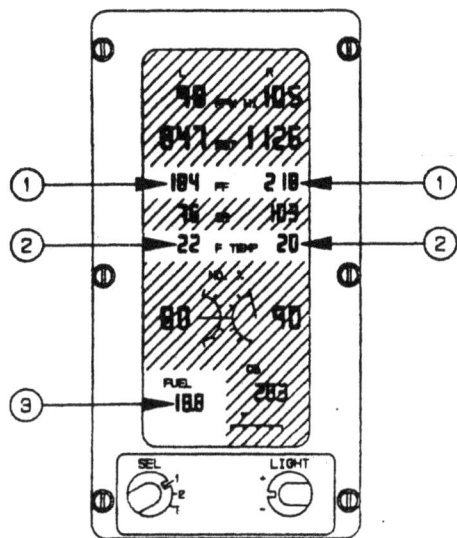

1. FUEL FLOW (L AND R)
2. FUEL TEMPERATURE (L AND R)
3. TOTAL FUEL QUANTITY

A22-052

*Figure 1-21*

takes manual control of the component and is indicated by a box around the component legend (figure 1-23). Successive pressing of the selected button cycles the component manually through all operational states.

Manual settings stay as selected as long as the manual fuel display is selected. If the manual fuel display is deselected, the manual settings remain for thirty seconds then automatically revert to automatic operation. The manual fuel display can be reselected within thirty seconds to maintain the manual settings.

Control of an individual component is returned to automatic by cycling through all operational states. The reset button labeled RSET or the control stick paddle switch returns all components to automatic control. Component states when a button is pressed are:

> O - Open,
> CL - Closed,
> ON - On,
> W - Wing,
> F - Feed,
> OFF - Off.

The manual operations available are:

| | |
|---|---|
| LW | Controls the refuel shutoff valve in the left wing. |
| LF | Controls the left feed tank refuel/transfer shutoff valve. |
| CONT | Transfers to automatic control. |
| RF | Controls the right feed tank refuel/transfer shutoff valve. |
| RW | Controls the refuel shutoff valve in the right wing. |
| RWP | Controls the right wing transfer pump to transfer fuel from the right wing tank to the engine feed tanks. |
| RSET | Returns manual settings to automatic settings. |
| RCD | Records the current configuration. |
| M | Return to main menu. |
| RDIV | Controls right fuel diverter valve to divert hot fuel from the right accessory cooling system to the right wing or right feed tank. |
| AFXFR | Controls the aft tank transfer inlet valve. |
| RDIS | Controls the redistribution valve to connect the transfer manifold to the refuel manifold (ground test). |
| CTXFR | Controls the center tank transfer inlet valve. |
| LDIV | Controls the left fuel diverter valve to divert hot recirculating fuel from the left accessory cooling system to the left wing tank or the left feed tank. |
| AFT | Controls the refuel shutoff valve in the aft tank. |
| CTR | Controls the refuel shutoff valve in the center tank. |
| XFP2 | Controls fuselage transfer pump number 2 to transfer fuel from the center and aft tanks to the engine feed tanks. |
| XFP1 | Controls fuselage transfer pump number 1 to transfer fuel from the center and aft tanks to the engine feed tanks. |

## FUEL DISPLAY (TYPICAL)

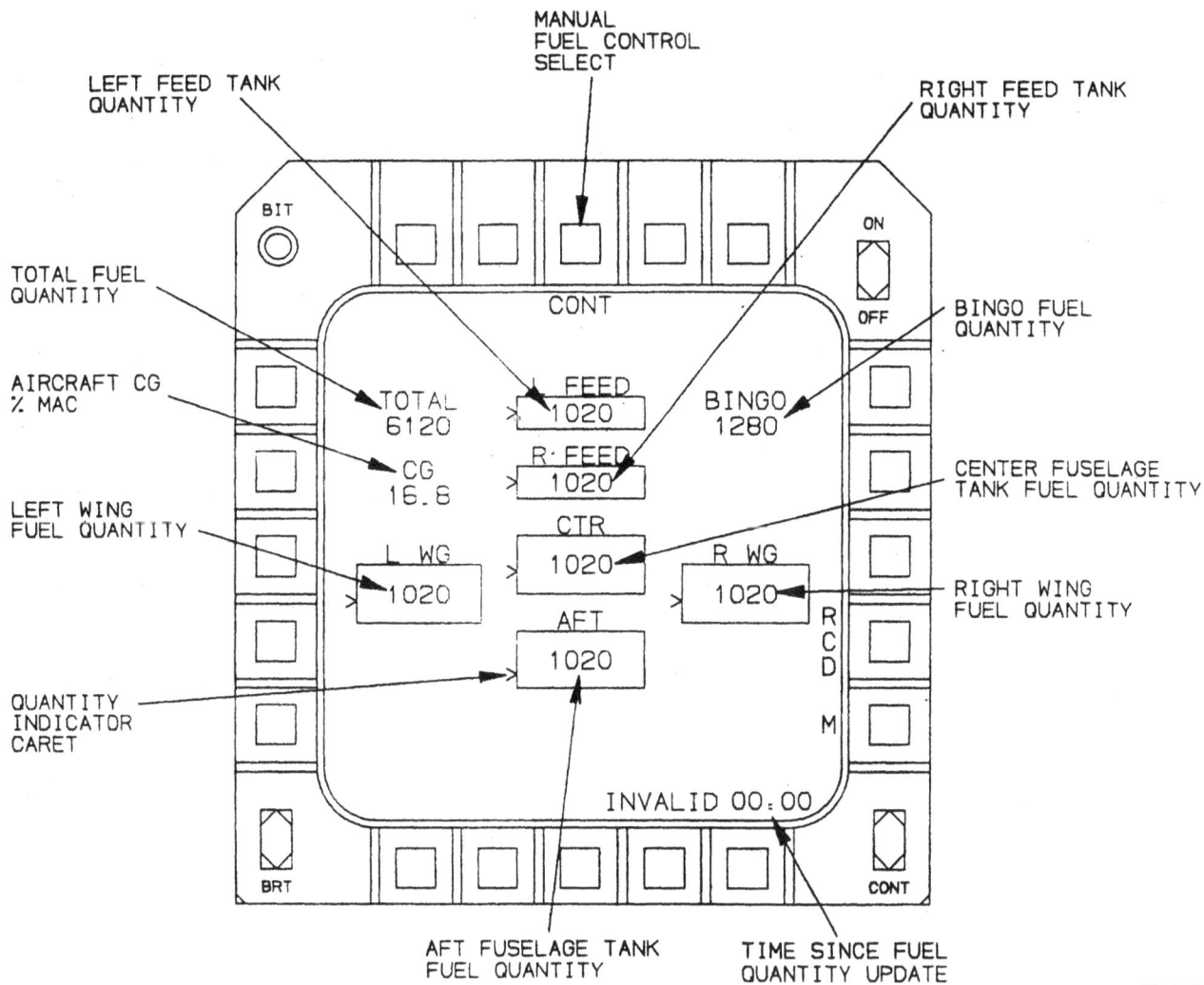

Figure 1-22

# MANUAL FUEL CONTROL DISPLAY

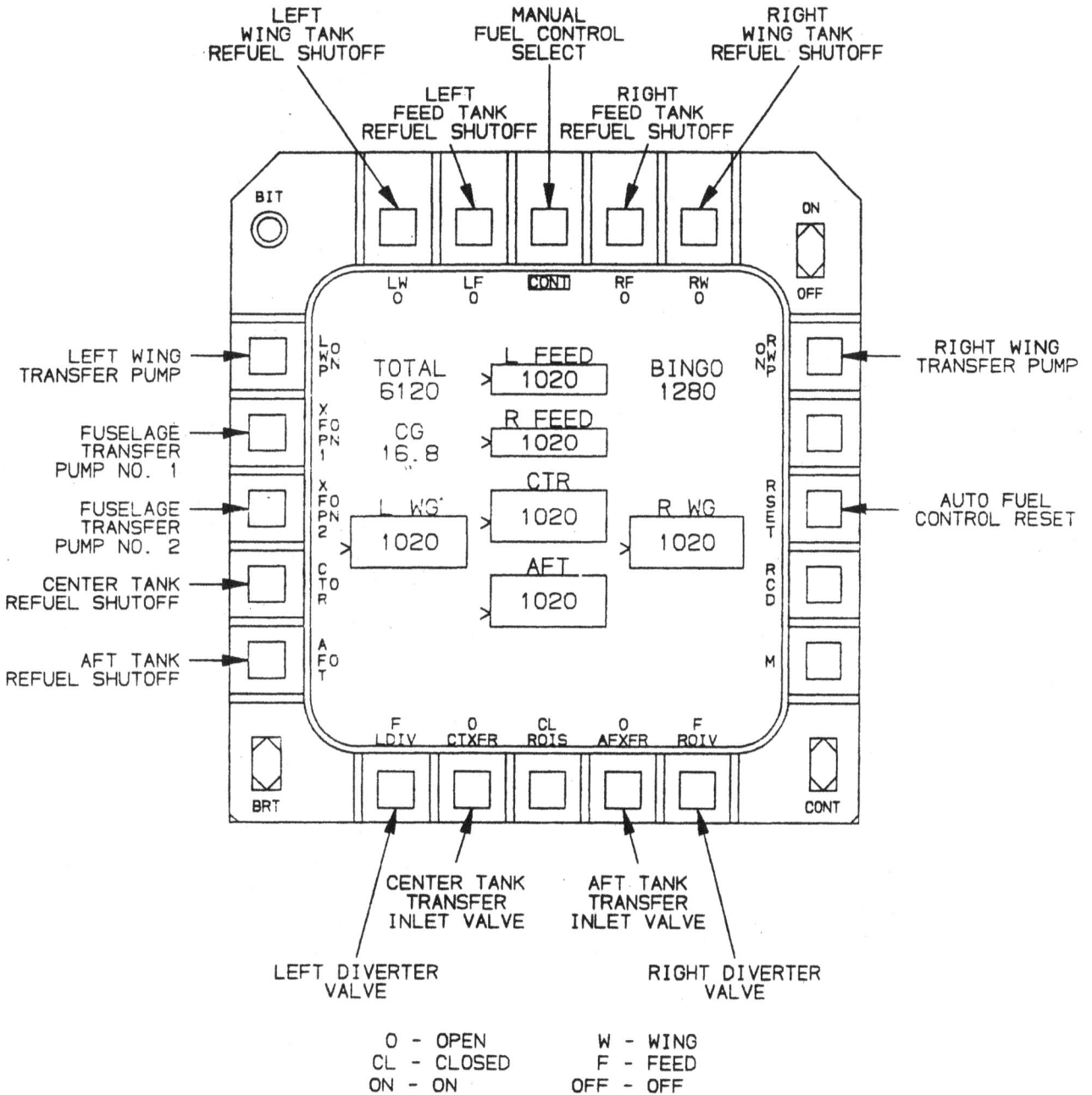

LEFT WING TANK REFUEL SHUTOFF

LEFT FEED TANK REFUEL SHUTOFF

MANUAL FUEL CONTROL SELECT

RIGHT FEED TANK REFUEL SHUTOFF

RIGHT WING TANK REFUEL SHUTOFF

LEFT WING TRANSFER PUMP

FUSELAGE TRANSFER PUMP NO. 1

FUSELAGE TRANSFER PUMP NO. 2

CENTER TANK REFUEL SHUTOFF

AFT TANK REFUEL SHUTOFF

RIGHT WING TRANSFER PUMP

AUTO FUEL CONTROL RESET

CENTER TANK TRANSFER INLET VALVE

AFT TANK TRANSFER INLET VALVE

LEFT DIVERTER VALVE

RIGHT DIVERTER VALVE

O - OPEN         W - WING
CL - CLOSED      F - FEED
ON - ON          OFF - OFF

BIT

LW 0    LF 0    CONT    RF 0    RW 0

ON / OFF

L W O P N
X F O P N 1
X F O P N 2
C T O R
A F O T

O R W N P
R S E T
R C D
M

TOTAL 6120

CG 16.8

L WG 1020

L FEED 1020

R FEED 1020

CTR 1020

AFT 1020

R WG 1020

BINGO 1280

F LDIV    O CTXFER    CL RDIS    O AFXFER    F RDIV

BRT          CONT

*Figure 1-23*

LWP      Controls left wing transfer pump to transfer fuel from the left wing tank to the engine feed tanks.

## FUEL CAUTIONS AND ADVISORIES

Cautions and advisories appear on the MPCD or HUD and are accompanied by a caution light or voice message.

### Fuel Cautions

The fuel cautions triggering a caution light are:

FUEL ASYM      Fuel imbalance (500 pounds) between left and right wings.

FUEL FAIL      Fuel management computer failure or loss of two quantity gauging probe in same tank.

FUEL HOT      High fuel temperature.

FUEL LOW      Low fuel in either main feed tank. Indication accompanied by the voice message *WARNING, FUEL LOW*

L (R) BST PUMP      Boost pump pressure low.

TANK PRESS      Fuel tank pressure low.

### Fuel Advisories

The MPCD fuel advisories are:

BINGO      Fuel quantity at preset level. Accompanied by voice warning message *BINGO FUEL* .

## AIR REFUELING SYSTEM

The air refueling system provides the means for the aircraft to receive fuel during flight. The air refueling manifold and the transfer manifold are the same manifold. All tanks may be air refueled. The system consists of the air refueling receptacle and slipway, signal amplifier, and the controls/indicators. A universal aerial refueling receptacle slipway installation modified with a curved door to conform to the aircraft moldline is used.

### Air Refueling Receptacle

The air refueling receptacle and latch mechanism are hydraulically operated and spring-loaded to open if hydraulic power is lost. The receptacle cannot be opened if electrical power is lost.

### Air Refueling Signal Amplifier

The air refueling signal amplifier receives signals from the coil and places the air refueling system in the ready, latched, or disconnect mode. The induction coil is recessed into the receptacle. A coil in the tanker system couples with the induction coil and signals boom latching between the tanker and receiver systems.

### Air Refueling Controls and Indicators

The air refueling switch is on the left console (figure 1-24).

SLIPWAY ORIDE      Opens the air refueling receptacle and enables manual boom latching. Only the receiver can trigger a disconnect.

OPEN      Opens the air refueling receptacle and enables normal air refueling operation.

CLOSE      Closes the air refueling receptacle.

Positioning the air refueling switch to SLIPWAY ORIDE or OPEN causes the advisory ARR RDY to appear on the HUD and the MPCD when the air refueling receptacle is open. The fuel tanks remain pressurized. The advisory goes out when the boom is engaged.

### Air Refueling Automatic Operation

Selection of tanks to be filled is automatic and fuel is shutoff as tanks become full.

### Air Refueling Manual Operation

The tanks to be filled must be manually selected on the fuel display. Aircraft CG must be controlled and monitored. Refueling can be accomplished in minimum time by filling all tanks simultaneously.

Level control valves automatically close the tank refueling valves when the tanks are full. Closure of the fuel valves causes a fuel line pressure surge and could cause a pressure disconnect.

### Air Refueling Distribution

The air refueling distribution system is the same as the ground refueling distribution system. The air refueling time for a 90% fuel load is approximately 8 minutes.

## AIR REFUELING SWITCH

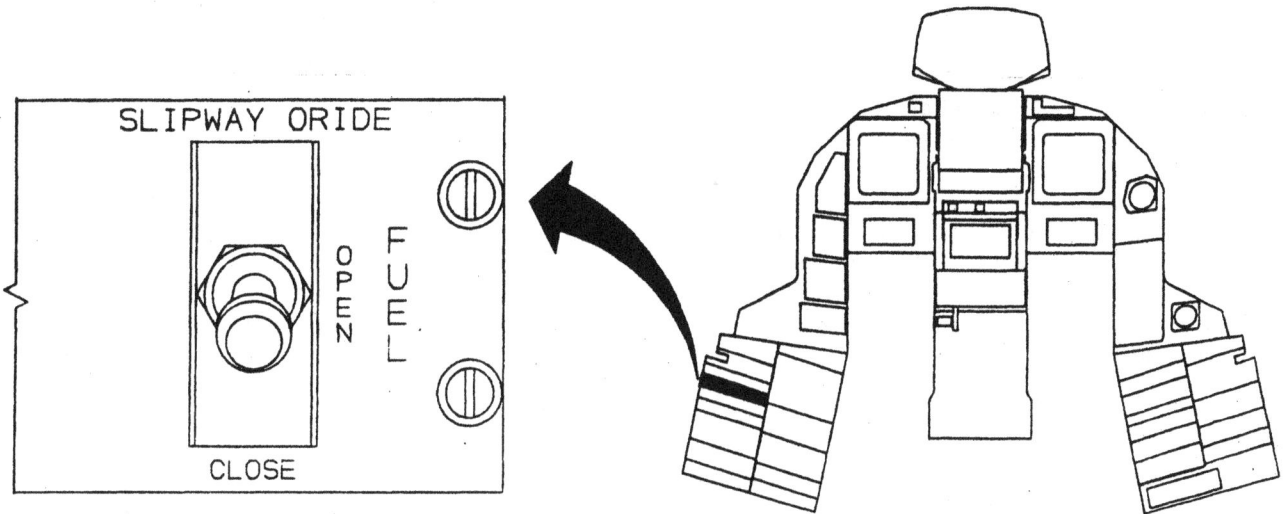

A23-054

Figure 1-24

### Air Refueling Termination

Air refueling can be terminated by:

- Tanks filled to 99 percent of volume.
- The air refueling disconnect switch.
- Tanker initiated disconnect.
- Tanker boom movement exceeding the boom envelope.
- Pressure disconnect.

A disconnect signal can be manually initiated by the receiver or the tanker boom operator. A brute force disconnect can be accomplished if a disconnect signal cannot be made. The advisory ARR RDY on the HUD and MPCD comes on when the boom disconnects.

### Air Refueling Advisory

Advisory appears on MPCD and HUD.

ARR RDY        Air refueling receptacle is open.

# ELECTRICAL SYSTEM

The electrical power system consists of two main ac generators, an emergency generator, two pairs of essential transformer-rectifiers, two vehicle management system transformer-rectifiers, a utility battery, an emergency battery, two vehicle management system (VMS) batteries, and ac and dc power distribution (bus) systems. External ac power can be applied to power the bus systems on the ground.

## AC ELECTRICAL POWER

The two main 40 kva ac generators are the primary source of electrical power. The two generators are connected for split bus nonsynchronized operation (figure 1-25). With both generators operating each generator supplies power independently to certain aircraft buses. If one generator fails and drops off the line, the remaining generator provides power to the essential ac buses, main ac bus 1 and the essential and main dc buses. The main ac bus 2 drops off line during single main generator operation. Current sensors are provided to prevent a fault in one generator system from shutting down both generators. Each generator activates automatically when the respective generator switch is in the position ON, and the generator connects to the proper buses when voltage and frequency are within prescribed limits. A protection system within the generator control unit protects against equipment damage due to undervoltage, overvoltage, over and under frequency, and feeder faults. If a malfunction occurs, the generator control unit removes the affected generator from the buses. The respective generator switch must be cycled to RESET and back to ON to bring the generator back on the line after a fault or out-of-tolerance condition clears. The generators may be reset as many times as necessary. A generator is removed from the buses by placing the generator switch OFF. Cautions L GEN or R GEN appear on the MPCD if the respective generator is off line.

### Generator Switches

The two main generator lever lock switches (figure 1-26) are labeled L GEN and R GEN respectively.

| | |
|---|---|
| ON | Generator automatically powers buses when running. |
| OFF | Generator off. |
| RESET (Momentary) | Resets generator. |

## Emergency Generator

If both main generators are off or failed, emergency ac power is supplied by a 15 kva generator driven by the emergency power unit (EPU) 1. The emergency generator powers the right and left essential ac buses and all six transformer rectifiers to supply power to the essential dc buses, the right main dc bus, and to the VMS buses 1 and 2. Emergency power unit operation appears on the MPCD. In emergency generator only operation the left and right main ac buses 1 and 2, and the left main dc bus are not powered. A protection system within the emergency generator control unit protects against equipment damage due to overvoltage or under-frequency. If a malfunction occurs, the emergency generator control unit removes the emergency generator from the buses and give associated caution E GEN MALF. The emergency generator switch must be cycled to RESET and back to ON to bring the emergency generator back on line after a fault or out-of-tolerance condition clears.

### Emergency Generator Switch

The lever lock emergency generator switch has two positions and is labeled EMERG GEN.

| | |
|---|---|
| ON | Generator automatically powers essential ac buses when running. |
| RESET (Momentary) | Resets emergency generator. |

The aircraft can be configured to operate the EPU with bleed air, refer to flight test card to select test configuration.

## DC ELECTRICAL POWER

Six transformer-rectifiers and four batteries are provided (figure 1-27). The output of the left transformer-rectifiers 1 and 2 and the right transformer-rectifiers 1 and 2 power the respective left and right essential dc buses. The output of the VMS 1 and 2 transformer-rectifiers power the respective VMS 1 and 2 dc buses. Either main generator or emergency generator can power all six transformer-rectifiers. Protection is provided so that a short in any transformer-rectifier does not affect another. The caution XFMR RECT on the MPCD indicates one or more transformer-rectifier have failed. The top level BIT display shows the failed transformer rectifier. The utility battery can power the utility bus, the essential dc buses, and in flight the VMS 1 bus. In flight the emergency battery can power the essential

## AC POWER SYSTEM

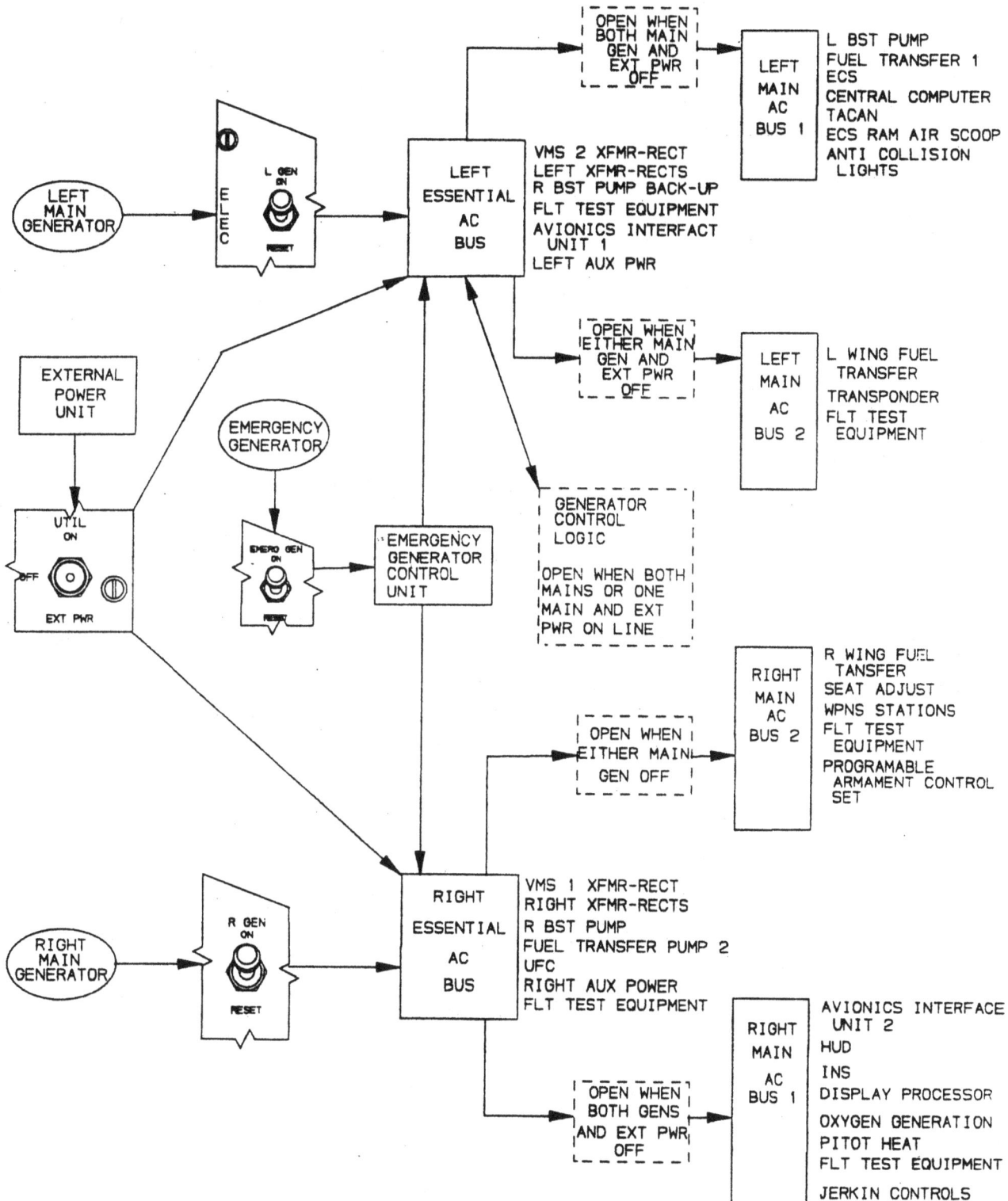

LEFT MAIN GENERATOR

ELEC — L GEN ON / RESET

LEFT ESSENTIAL AC BUS
- VMS 2 XFMR-RECT
- LEFT XFMR-RECTS
- R BST PUMP BACK-UP
- FLT TEST EQUIPMENT
- AVIONICS INTERFACT UNIT 1
- LEFT AUX PWR

OPEN WHEN BOTH MAIN GEN AND EXT PWR OFF

LEFT MAIN AC BUS 1
- L BST PUMP
- FUEL TRANSFER 1
- ECS
- CENTRAL COMPUTER
- TACAN
- ECS RAM AIR SCOOP
- ANTI COLLISION LIGHTS

OPEN WHEN EITHER MAIN GEN AND EXT PWR OFF

LEFT MAIN AC BUS 2
- L WING FUEL TRANSFER
- TRANSPONDER
- FLT TEST EQUIPMENT

EXTERNAL POWER UNIT

UTIL ON / OFF — EXT PWR

EMERGENCY GENERATOR

EMERG GEN ON / RESET

EMERGENCY GENERATOR CONTROL UNIT

GENERATOR CONTROL LOGIC

OPEN WHEN BOTH MAINS OR ONE MAIN AND EXT PWR ON LINE

RIGHT MAIN AC BUS 2
- R WING FUEL TANSFER
- SEAT ADJUST
- WPNS STATIONS
- FLT TEST EQUIPMENT
- PROGRAMABLE ARMAMENT CONTROL SET

OPEN WHEN EITHER MAIN GEN OFF

RIGHT MAIN GENERATOR

R GEN ON / RESET

RIGHT ESSENTIAL AC BUS
- VMS 1 XFMR-RECT
- RIGHT XFMR-RECTS
- R BST PUMP
- FUEL TRANSFER PUMP 2
- UFC
- RIGHT AUX POWER
- FLT TEST EQUIPMENT

OPEN WHEN BOTH GENS AND EXT PWR OFF

RIGHT MAIN AC BUS 1
- AVIONICS INTERFACE UNIT 2
- HUD
- INS
- DISPLAY PROCESSOR
- OXYGEN GENERATION
- PITOT HEAT
- FLT TEST EQUIPMENT
- JERKIN CONTROLS

A23-061

**Figure 1-25**

# *ELECTRICAL CONTROLS*

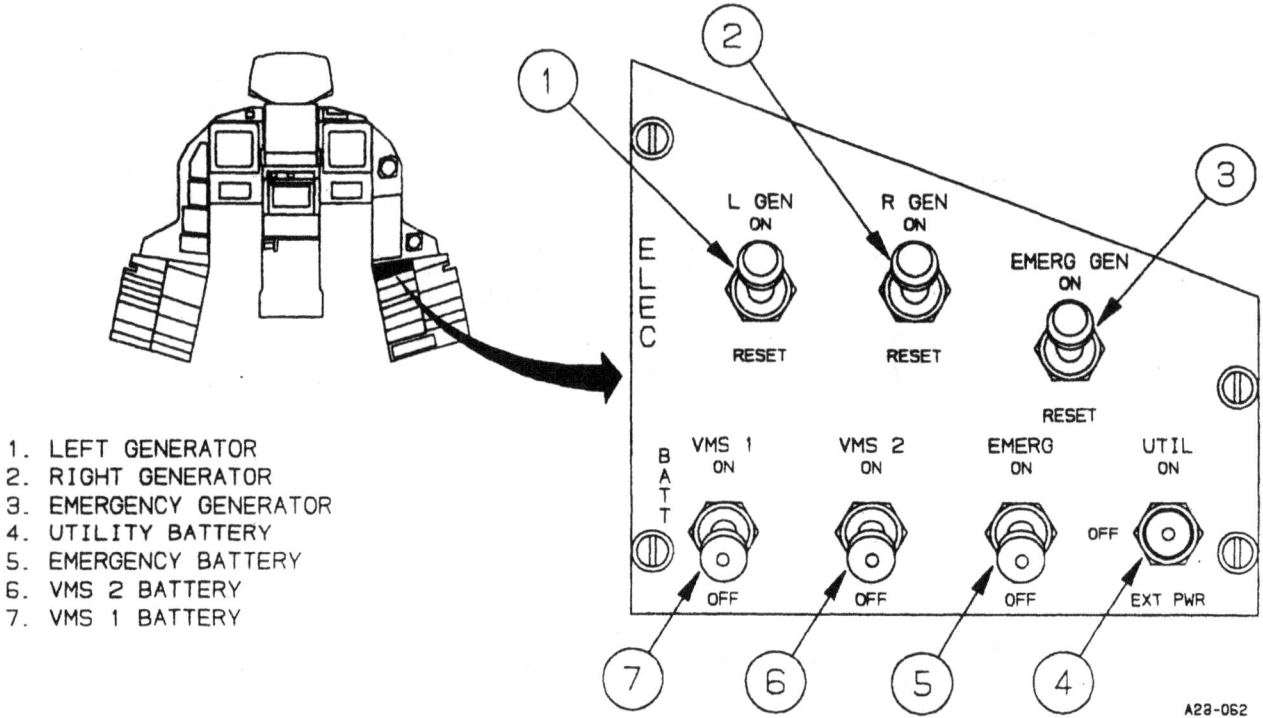

1. LEFT GENERATOR
2. RIGHT GENERATOR
3. EMERGENCY GENERATOR
4. UTILITY BATTERY
5. EMERGENCY BATTERY
6. VMS 2 BATTERY
7. VMS 1 BATTERY

A23-062

*Figure 1-26*

dc buses and the VMS 1 and 2 buses. The VMS 1 and 2 batteries power the respective VMS buses, when required. On the ground without ac power the VMS battery switches should be turned off or rapid depletion of the VMS batteries occurs.

## Batteries

The four batteries are controlled by switches lever locked to the ON positions. Pull out to move to the OFF position.

With the exception of the emergency battery, all batteries recharge with the battery switch ON and ac power available. The emergency battery does not recharge on the ground.

## Utility Battery Switch

The utility battery switch has three positions.

| | |
|---|---|
| ON | Battery powers utility, right, and left essential and VMS 2 dc buses if otherwise unpowered. In flight connects also to VMS 1 bus. |
| OFF | Battery disconnected. |

| | |
|---|---|
| EXT PWR (momentary) | Enables external power, when connected, to power the aircraft buses if both aircraft generators are not on line. |

## Emergency Battery Switch

The emergency battery is protected by a weight-on-wheels switch and only works in flight.

| | |
|---|---|
| ON | Battery powers left and right essential, VMS 1, and VMS 2 dc buses in flight if otherwise unpowered. |
| OFF | Battery is disconnected. |

## VMS 1 Battery Switch

The VMS 1 battery switch has two positions.

| | |
|---|---|
| ON | Battery powers VMS 1 bus if otherwise unpowered. |
| OFF | Battery is disconnected. |

## VMS 2 Battery Switch

The VMS 2 battery switch has two positions.

## DC POWER SYSTEM

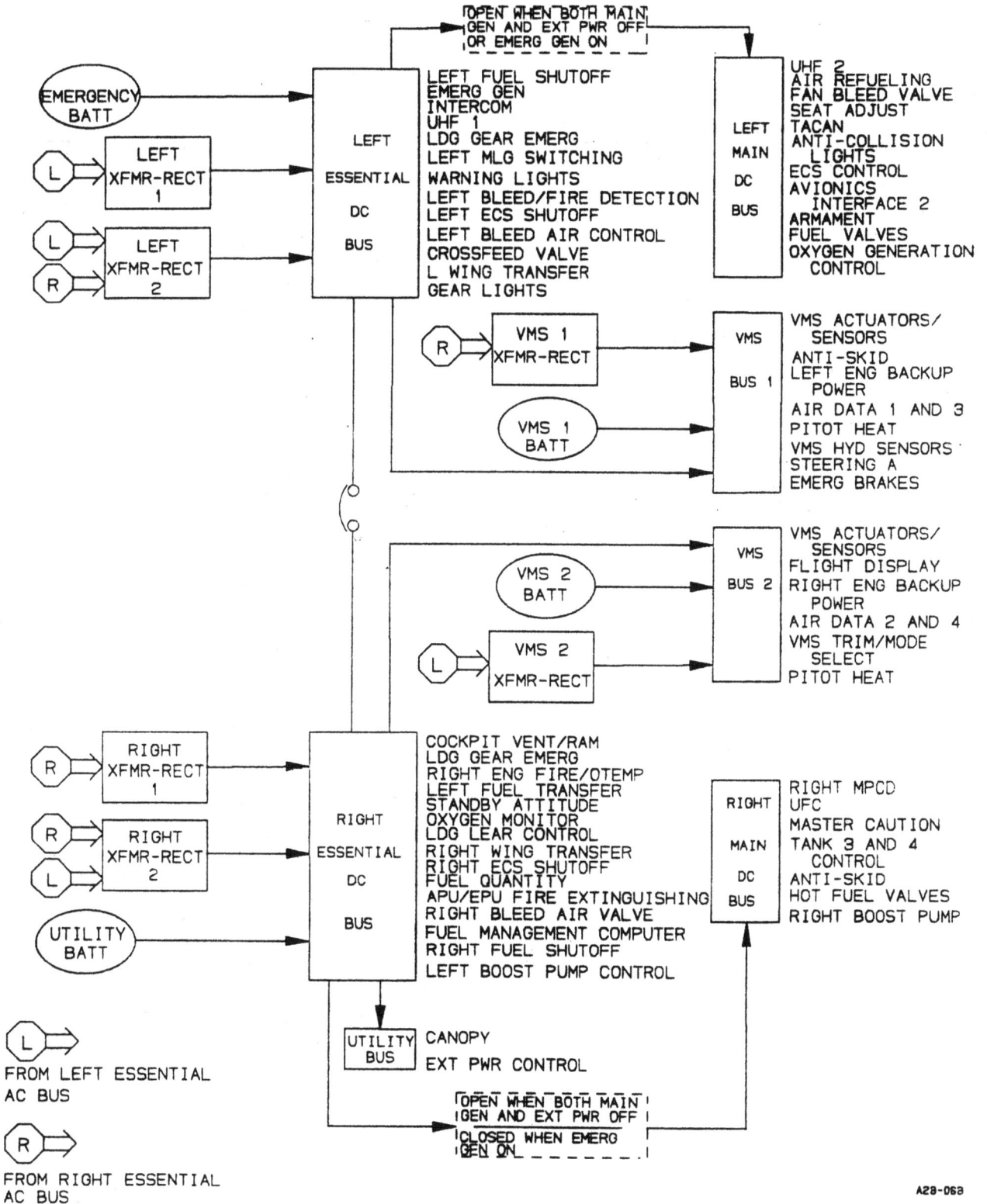

OPEN WHEN BOTH MAIN
GEN AND EXT PWR OFF
OR EMERG GEN ON

EMERGENCY BATT

LEFT XFMR-RECT 1

LEFT XFMR-RECT 2

**LEFT ESSENTIAL DC BUS**

LEFT FUEL SHUTOFF
EMERG GEN
INTERCOM
UHF 1
LDG GEAR EMERG
LEFT MLG SWITCHING
WARNING LIGHTS
LEFT BLEED/FIRE DETECTION
LEFT ECS SHUTOFF
LEFT BLEED AIR CONTROL
CROSSFEED VALVE
L WING TRANSFER
GEAR LIGHTS

**LEFT MAIN DC BUS**

UHF 2
AIR REFUELING
FAN BLEED VALVE
SEAT ADJUST
TACAN
ANTI-COLLISION
    LIGHTS
ECS CONTROL
AVIONICS
    INTERFACE 2
ARMAMENT
FUEL VALVES
OXYGEN GENERATION
    CONTROL

VMS 1 XFMR-RECT

VMS 1 BATT

**VMS BUS 1**

VMS ACTUATORS/
    SENSORS
ANTI-SKID
LEFT ENG BACKUP
    POWER
AIR DATA 1 AND 3
PITOT HEAT
VMS HYD SENSORS
STEERING A
EMERG BRAKES

VMS 2 BATT

VMS 2 XFMR-RECT

**VMS BUS 2**

VMS ACTUATORS/
    SENSORS
FLIGHT DISPLAY
RIGHT ENG BACKUP
    POWER
AIR DATA 2 AND 4
VMS TRIM/MODE
    SELECT
PITOT HEAT

RIGHT XFMR-RECT 1

RIGHT XFMR-RECT 2

UTILITY BATT

**RIGHT ESSENTIAL DC BUS**

COCKPIT VENT/RAM
LDG GEAR EMERG
RIGHT ENG FIRE/OTEMP
LEFT FUEL TRANSFER
STANDBY ATTITUDE
OXYGEN MONITOR
LDG LEAR CONTROL
RIGHT WING TRANSFER
RIGHT ECS SHUTOFF
FUEL QUANTITY
APU/EPU FIRE EXTINGUISHING
RIGHT BLEED AIR VALVE
FUEL MANAGEMENT COMPUTER
RIGHT FUEL SHUTOFF
LEFT BOOST PUMP CONTROL

**RIGHT MAIN DC BUS**

RIGHT MPCD
UFC
MASTER CAUTION
TANK 3 AND 4
    CONTROL
ANTI-SKID
HOT FUEL VALVES
RIGHT BOOST PUMP

**UTILITY BUS**

CANOPY
EXT PWR CONTROL

OPEN WHEN BOTH MAIN
GEN AND EXT PWR OFF
CLOSED WHEN EMERG
GEN ON

L FROM LEFT ESSENTIAL AC BUS

R FROM RIGHT ESSENTIAL AC BUS

A28-063

*Figure 1-27*

ON            Battery powers VMS 2 bus if otherwise unpowered.

OFF          Battery is disconnected.

## ELECTRICAL CAUTIONS

**Electrical system cautions appear on the MPCD and turn on the master caution light.**

L GEN          Left main generator failed or disconnected.

R GEN          Right main generator failed or disconnected.

BATTERY
- Emergency, Utility, VMS 1, or VMS 2 battery failure.
- A battery switch is OFF.

XFMR RECT     Single or multiple transformer-rectifier failure. BIT identifies specific failure.

E GEN MAL     Emergency generator not producing power. Caution appears only if ac power available.

## EMERGENCY POWER SYSTEM

The emergency power system provides emergency electrical power and utility hydraulic pressure. The system automatically activates in flight upon command from the VMS for dual engine flameout, dual main generator failure, or loss of utility hydraulics and a flight control hydraulic system. The system can be activated manually. The system draws electrical power from the left essential DC bus. Main components of the emergency power system are

- Two emergency power units (EPU),
- Two hydraulic pumps,
- One generator,
- Two hydrazine tanks,
- One nitrogen tank,
- Cockpit controls and indicators.

EPU 1 drives an emergency generator and a hydraulic pump. EPU 2 drives a hydraulic pump. The EPU are located in the [1] [2] APU bay or the [2] weapons bay.

Hydrazine is forced into a decomposition chamber. The reaction gases spin a turbine gearbox and power the generator and hydraulic pumps. Exhaust vents overboard [1] [2] out the top of the APU bay or [2] out the bottom of the weapons bay door. The flammable exhaust can have temperatures up to 900 degrees Fahrenheit. Exhaust gases have an ammonia odor and are irritating to eyes and nose.

A ground safety switch in the right wheel well disables the emergency power system on the ground.

### EPU Selector

A four position rotary EPU selector controls EPU operation, figure 1-28. The selector is pushed in and rotated from AUTO to ON.

| | |
|---|---|
| OFF | Both EPU off. |
| BLD | [1] No function.<br>[2] Bleed air powers EPU 1 (when bleed air plumbing is installed for high angle of attack testing). |
| AUTO | EPU starts automatically upon command from the VMS. |
| ON | Both EPU on |

### EPU OPERATION

Dual engine flameout or loss of the utility hydraulic system and either flight control hydraulic system starts both EPU. Loss of both generators starts only

## EPU CONTROL

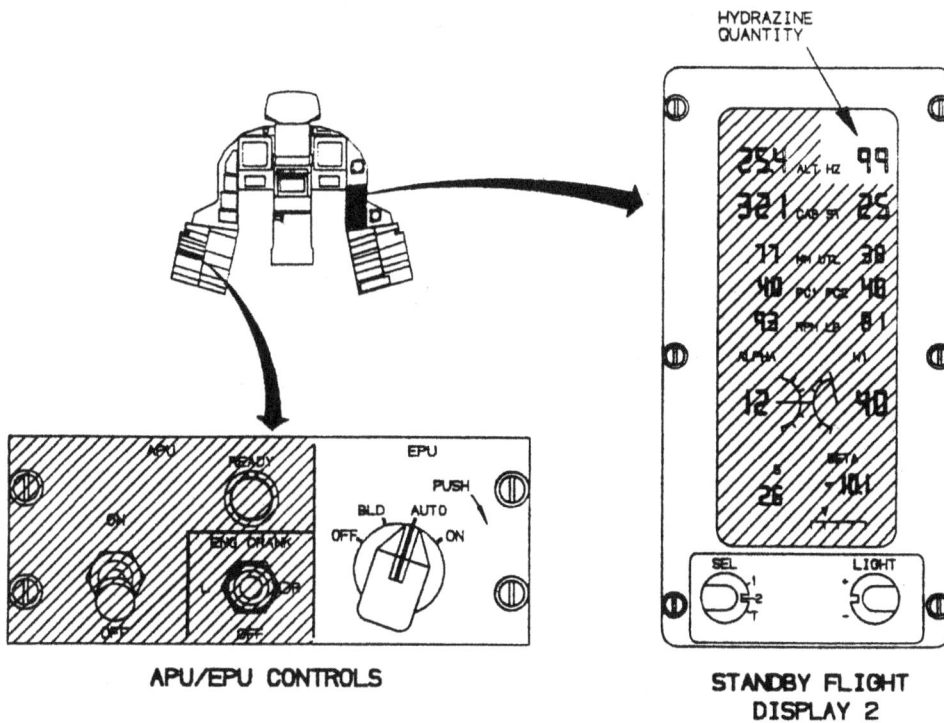

APU/EPU CONTROLS

STANDBY FLIGHT
DISPLAY 2

HYDRAZINE
QUANTITY

Figure 1-28

EPU 1. The system can operate for approximately 9 minutes using hydrazine. Hydrazine remaining in percent of full, appears on the standby flight display 2.

## EPU BLEED AIR OPERATION [2]

When bleed air plumbing is installed for high angle of attack testing, EPU is located in the APU bay. EPU operation uses engine bleed air and/or hydrazine to spin the turbine/gearbox. When bleed air flow is insufficient, hydrazine augmentation occurs automatically.

## EPU CAUTIONS

EPU cautions appear on the MPCD:

E GEN MALF      Electrical power from EPU 1 is overvoltage or underfrequency.

EPU NOT ARM      EPU ground safety switch in safe or EPU selector is in OFF.

HYDRAZINE      EPU operating using hydrazine.

## EPU ADVISORIES

EPU advisories appear on the MPCD:

EPU GEN      EPU 1 is operating.

EPU HYD      EPU 2 is operating.

# HYDRAULIC SYSTEM

Aircraft hydraulic power is supplied by three independent 4000 psi hydraulic systems (figure 1-29). The systems are power control system one (PC-1), power control system two (PC-2), and the utility system. Any one of the three systems can power the flight controls.

The power control systems supply hydraulic pressure to the actuators of all flight control surfaces. AMAD mounted hydraulic pumps on each engine pressurize the hydraulic systems. Each AMAD powers one PC pump and one utility pump. Hydraulic pumps are identical and operate continuously while engines are running.

## POWER CONTROL SYSTEMS

Flight controls are powered by PC systems 1 and 2. The left engine AMAD drives the PC-1 pump and the right engine drives the PC-2 pump. The PC-1 pump supplies pressure to the left leading edge flaps, left and right inboard and outboard flaperons and both tail surfaces. PC-2 supplies pressure to the right leading edge flaps, left and right inboard and outboard flaperons and both tail surfaces. If either PC-1 or PC-2 fails, switching valves allow the utility system to power the failed system.

## UTILITY HYDRAULIC SYSTEM

The utility hydraulic system acts as a backup system to power the flight controls and is pressurized by two hydraulic pumps; one on each AMAD. The right utility system pressure is approximately 100 psi lower than the left due to system preload pressure. Both pumps deliver pressure to a common manifold later divided into two circuits (A and B). Utility circuit A primarily pressurizes the emergency accumulator. Utility circuit B supplies hydraulic pressure to:

- landing gear,
- nose wheel steering,
- wheel brakes,
- weapons bay doors,
- inlet bleed doors (right and left),
- air refueling receptacle.

If utility circuit B fails, the emergency accumulator provides limited backup hydraulic pressure for emergency operation of landing gear extension, wheel braking, and nosewheel steering. Utility circuit A can recharge a fully depleted emergency accumulator to 205 cubic inches in approximately 30 seconds.

If utility circuit A fails, the emergency accumulator provides:

- Landing gear extension -- 140 cubic inches.
- Wheel brake application -- 1.5 cubic inches.
- Nosewheel steering application --
  20 degree cycle -- 6 inches.
  45 degree cycle -- 11 inches.

Anti-skid protection is not provided with the utility hydraulic circuit B inoperative.

### Reservoir Level Sensing Valves

The utility hydraulic fluid reservoir has two reservoir level sensing (RLS) valves to isolate leaks in the respective circuits. Leak isolation logic is reservoir fluid level dependent. If a leak develops in a utility circuit, lowering the reservoir level, the RLS valve on circuit A closes first. If the leak continues, (leak in circuit B), the RLS valve on circuit B closes and the RLS valve on circuit A reopens.

### RESERVOIRS

Each hydraulic system has one reservoir. The reservoirs have remote level indicators for ground servicing of the systems. If a leak depletes reservoir fluid below a preset level, a signal is sent to the VMS computer. The caution FCS 1ST could be the first indication of PC system failure.

### HYDRAULIC FLUID HEAT EXCHANGERS

Two airframe mounted AMAD equipment oil heat exchangers cool the hydraulic fluid. Heat from the hydraulic fluid in all three systems is transferred to engine feed fuel.

### EMERGENCY HYDRAULIC PUMPS

Two emergency power unit (EPU) hydraulic pumps supply power to the utility hydraulic system if both engines fail, or the utility hydraulic system and either PC system fails. Emergency hydraulic system pressure is 3000 psi. Flight control movement could be less responsive while operating on emergency power unit hydraulic pumps.

### HYDRAULIC INDICATIONS

Hydraulic pressure is presented on the standby flight display 2 (figure 1-30). System pressure appears as a two digit number. For example, a readout display of 38 means 3800 psi. Utility pressure appears to the right of the legend UTL. The highest utility (normally utility 2) pressure is shown.

### HYDRAULIC CAUTIONS

Hydraulic cautions appear on the MPCD and trigger a caution light:

## *HYDRAULIC SYSTEM*

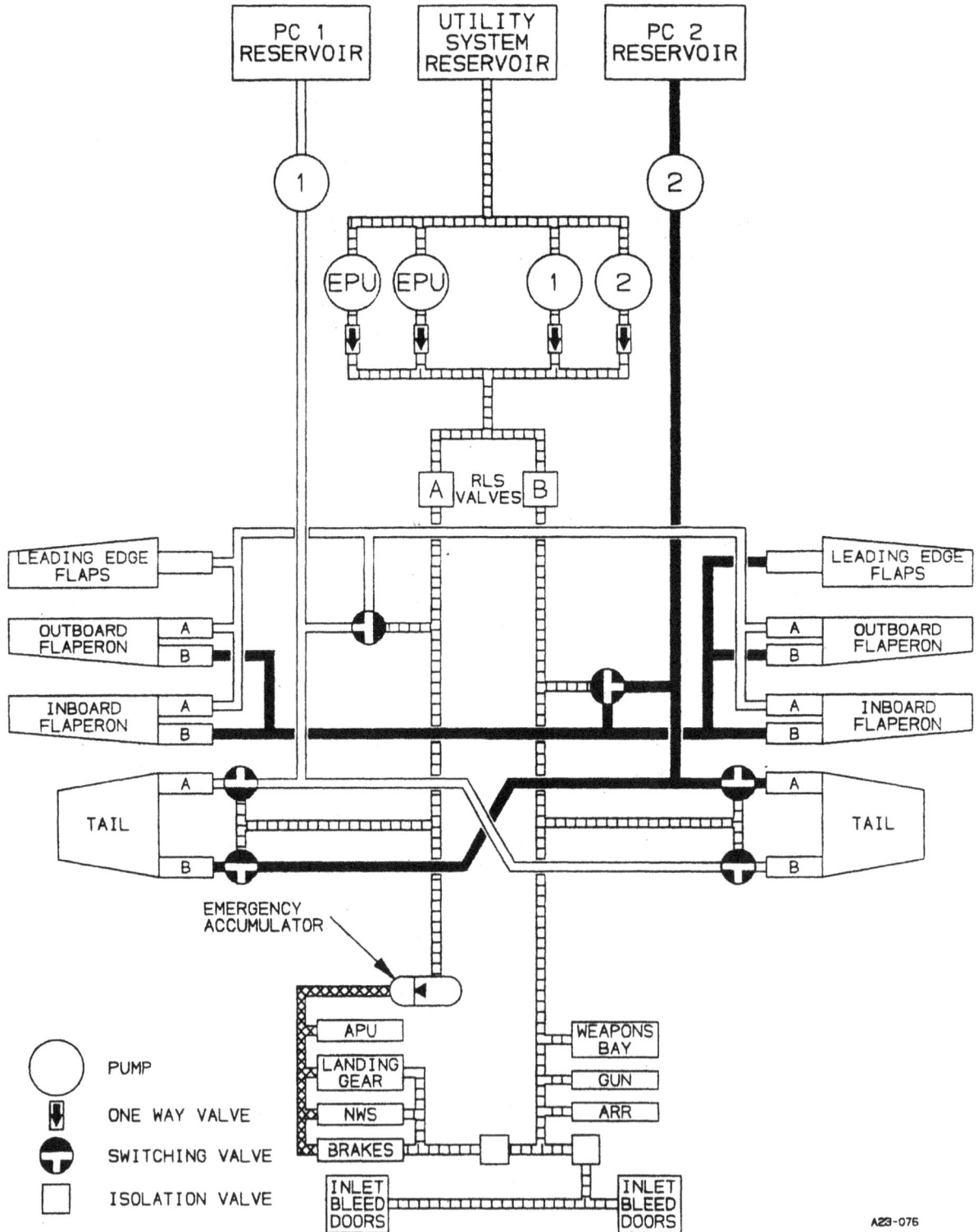

Figure 1-29

## STANDBY FLIGHT DISPLAY 2 -- HYDRAULIC

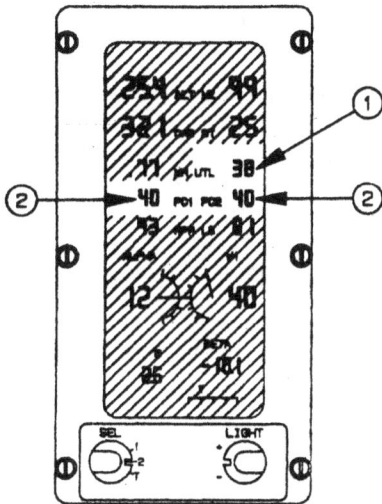

1. UTILITY HYDRAULIC PRESSURE (UTL)
2. FLIGHT CONTROL HYDRAULIC PRESSURE (L AND R)

A28-077

*Figure 1-30*

| | |
|---|---|
| PC-1 | Flight control hydraulic system 1 pressure is below 1700 psi. |
| PC-2 | Flight control hydraulic system 2 pressure is below 1700 psi. |
| UTIL 1 | Left utility hydraulic pump pressure below 1700 psi. |
| UTIL 2 | Right utility hydraulic pump pressure below 1700 psi. |
| UTIL A | Utility A circuit pressure is below 1700 psi. |
| UTIL B | Utility B circuit pressure is below 1700 psi. |
| APU ACCUM | Emergency accumulator pressure is below 3400 psi. |

**UNCLASSIFIED**
**NORTHROP/MCAIR YF-23A**
**COMPETITION SENSITIVE**

# LANDING GEAR SYSTEM

The tricycle landing gear consists of a cantilevered nose gear and semi-levered main gear. The landing gear system includes brakes, antiskid protection, and nosewheel steering. With the landing gear up after takeoff, the utility hydraulic system isolation valve closes to isolate circuit B from the landing gear, nosewheel steering, and brakes

# LANDING GEAR

The landing gear and gear doors are hydraulically operated and electrically controlled and sequenced. Hydraulic power for extension and retraction is furnished by the utility hydraulic system. The emergency accumulator provides power for landing gear emergency extension.

Normal gear retraction is approximately 4.5 seconds. Landing gear and doors are retained in the retracted position by mechanical uplocks.

Normal gear extension is approximately 7.0 seconds. Main landing gear is maintained in the extended position by a drag brace held on-center by an internal locking actuator. Nose gear is maintained down and locked by an over-center drag brace and spring. Gear doors remain open after landing gear extension.

## LANDING GEAR CONTROL PANEL

### Gear Handle

The landing gear is controlled by a two-position, wheel-shaped handle (figure 1-31).

| | |
|---|---|
| DN | Extends the landing gear. |
| UP | Retracts the landing gear. |

### Gear Handle Warning Light

A red light in the gear handle comes on steady when the landing gear and/or gear doors are not in the position commanded by the gear handle. If the gear is not down and locked, the red light flashes and a tone sounds when the aircraft is below 7500 feet MSL at less than 175 KCAS and descending at more than 200 feet per minute. If the gear handle is up and the gear control circuit power fails, the warning light comes on but the tone does not sound.

### Gear Indicator Lights

The landing gear down indicator lights, labeled NOSE/LEFT/RIGHT, come on green when the respective landing gear is down and locked.

### Warning Tone Silencer.

A beeping gear warning tone is activated with the flashing gear warning red light. The warning tone silence button, labeled WARN TONE SIL, turns off the warning tone but the red light in the gear handle continues to flash.

## LANDING GEAR EMERGENCY EXTENSION

The landing gear can be extended by using the landing gear emergency system. The hydraulic emergency accumulator supplies hydraulic pressure to the landing gear door actuators, uplock actuators, and the main landing gear extend actuators. The nose gear free-falls to the extend position.

Landing gear emergency extension could take approximately 25-35 seconds.

### Landing Gear Emergency Extension Switch

A two-position, lever-locked to NORM, landing gear emergency extension switch provides an emergency means of extending the landing gear. When pulled out and placed to the position DOWN, emergency extension of the landing gear is initiated, regardless of landing gear handle position.

# BRAKES

The main landing gear wheels have full power brakes operated by toe action on the rudder pedals. Normal brake pressure is supplied by the utility hydraulic system. Emergency brake pressure is supplied by the hydraulic emergency accumulator. An antiskid system is incorporated with the brakes to prevent tire skid. A brake hold function is provided and controlled by the antiskid switch.

## BRAKE SWITCH

The brake system has a two-position switch to control the mode of braking.

| | |
|---|---|
| MECH | Provides direct mechanical linkage to the hydraulic brake control valve. Spring loaded to MECH. |
| BBW | Selects a computer assisted brake-by-wire mode that regulates brake pressure through the antiskid system. Electro-magnetically held in BBW. |

## BRAKE HOLD

The brake hold is engaged when both engines are in idle, both brake pedals are pressed over 50% deflection and the antiskid switch is moved to HOLD. The brake hold is disengaged when both throttles are

**NORTHROP/MCAIR YF-23A**
**COMPETITION SENSITIVE**
**UNCLASSIFIED**

# LANDING GEAR SYSTEM CONTROLS AND INDICATORS

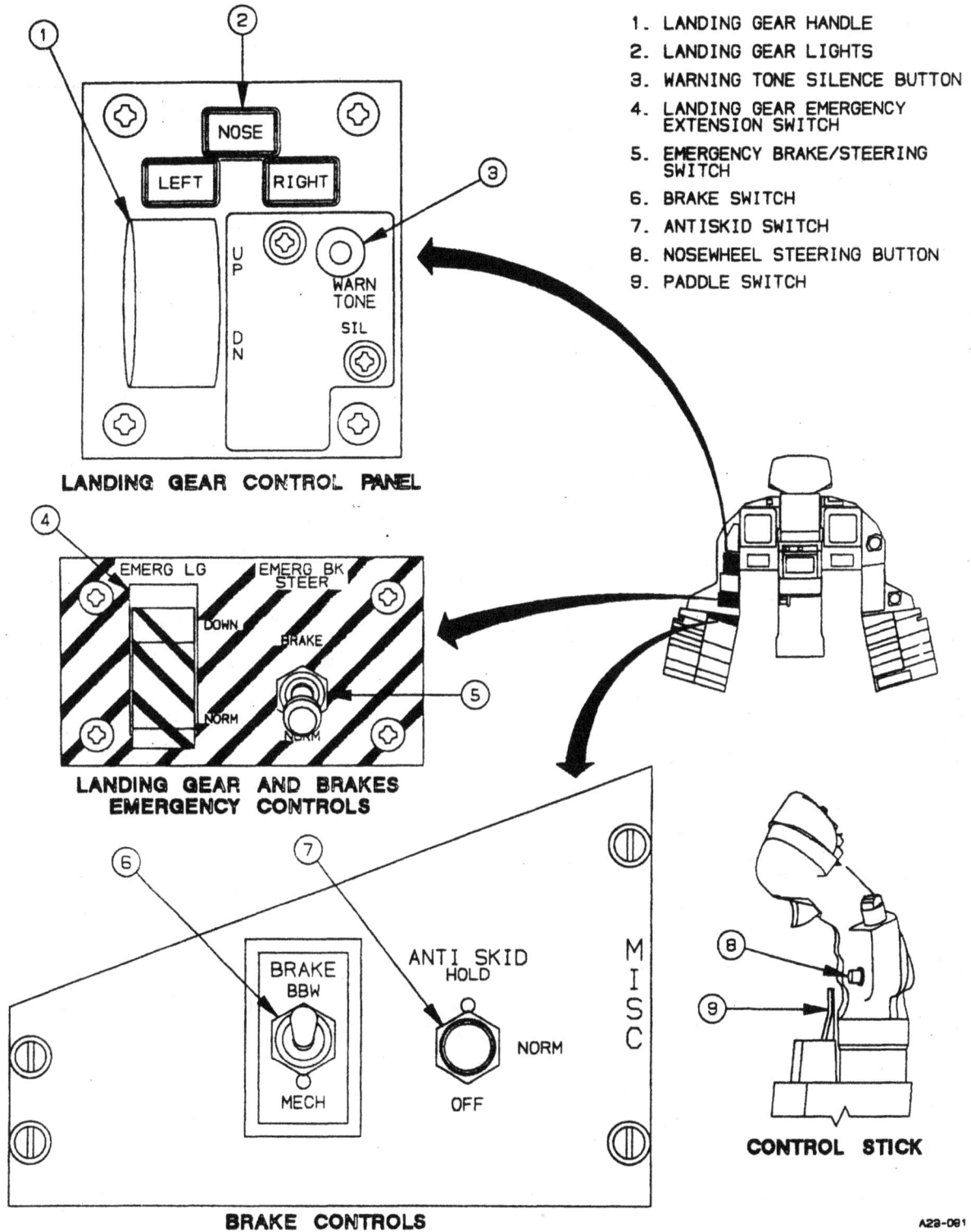

1. LANDING GEAR HANDLE
2. LANDING GEAR LIGHTS
3. WARNING TONE SILENCE BUTTON
4. LANDING GEAR EMERGENCY EXTENSION SWITCH
5. EMERGENCY BRAKE/STEERING SWITCH
6. BRAKE SWITCH
7. ANTISKID SWITCH
8. NOSEWHEEL STEERING BUTTON
9. PADDLE SWITCH

LANDING GEAR CONTROL PANEL

LANDING GEAR AND BRAKES EMERGENCY CONTROLS

BRAKE CONTROLS

CONTROL STICK

A23-091

*Figure 1-31*

moved above idle, the antiskid switch is moved out of the HOLD position, or both brakes are pressed over 60% deflection.

The brake hold should not be used after a braked stop that measures outside the normal range on the Brake Energy Limitations chart, Section III.

## EMERGENCY BRAKING

The emergency brake system is activated by the emergency brake switch. The system uses hydraulic accumulator pressure through independent hydraulic lines to provide braking without antiskid protection. After landing gear emergency extension, the emergency accumulator provides approximately 10 full brake applications and 7 full nosewheel steering deflections.

### Emergency Brake/Steering Switch

The two-position emergency brake/steering switch provides an alternate source of hydraulic pressure for the brakes and nosewheel steering system.

| | |
|---|---|
| NORM | Utility system B hydraulic pressure is used for normal braking and nosewheel steering. |
| BRAKE | Emergency hydraulic accumulator pressure is used for emergency braking and nosewheel steering. |

## ANTISKID SYSTEM

The antiskid system is electrically controlled by a three-position switch. With antiskid on, hydraulic pressure to the brakes is inhibited until main wheels spin up to 70 knots (or 3 seconds after weight on main landing gear, if spin-up sensor fails.

An antispin feature is provided to automatically stop the main wheels from spinning after takeoff when gear retraction is commanded.

Antiskid protection is not available below 35 knots.

### ANTISKID SWITCH

| | |
|---|---|
| OFF | No antiskid protection. |
| NORM | Antiskid system selected. |
| HOLD | Brake hold selected. |

## NOSEWHEEL STEERING SYSTEM

Nosewheel steering is electrically controlled by rudder pedal position and hydraulically powered by utility hydraulic circuit B. Nosewheel steering is energized after aircraft power-up by momentarily pressing the nosewheel steering button. Nosewheel steering operates with weight on the nose gear. Steering deflection is 20 degrees.

A maneuvering steering mode is provided for use when taxiing below 15 knots ground speed. Maneuvering mode provides up to 45 degrees maximum steering and is engaged by holding the nosewheel steering button. Use of maximum deflection nosewheel steering should be restricted to below five knots ground speed.

After landing, a built-in two second delay is provided before nosewheel steering responds to rudder pedal position.

Nosewheel steering is disengaged by holding the paddle switch pressed, allowing the nosewheel to free caster. The nosewheel steering actuator provides nosewheel shimmy damping with or without nosewheel steering engaged.

## NOSEWHEEL EMERGENCY STEERING

Nosewheel emergency steering is selected by the dual function switch labeled EMERG BK/STEER. The system uses emergency hydraulic accumulator pressure through independent hydraulic lines to power the steering actuator.

# LANDING GEAR SYSTEM CAUTIONS

| | |
|---|---|
| ANTISKID | Antiskid switch is off, an antiskid system malfunction, or emergency braking is selected. |
| HOLD BRAKE | Brake hold is selected but not engaged. |
| NWS | Nosewheel steering off, inoperative, or system fault detected. |

# LANDING GEAR SYSTEM ADVISORY

| | |
|---|---|
| BBW OFF (MPCD) | • Brake-by-wire mode disengaged by VMS for system fault. <br> • Antiskid system off. |

# ENVIRONMENTAL CONTROL SYSTEM

The environmental control system (ECS) provides conditioned air for:

- Cockpit pressurization and air conditioning,
- Windscreen defog,
- Canopy seal,
- Anti-G,
- Fuel tank pressurization,
- Oxygen generation,
- Avionics cooling.

The ECS operates with engine, APU, or ground cart bleed air. Ram air from the aux ram air scoop or from the engine inlet is the heat sink through the primary heat exchanger (figure 1-32).

## BLEED AIR SYSTEM

The bleed air ducts from each engine are joined after the primary pressure regulator valve and routed through the secondary pressure regulator valve to the primary heat exchanger. Warm air is routed to the ECS from the primary heat exchanger. Check valves prevent reverse flow.

### Bleed Air Control

Bleed air control is on the ECS panel (figure 1-33).

### BLEED AIR CONTROL

The control labeled BLEED AIR selects the bleed air source. The control is pulled to provide APU augmented airflow for ECS operation.

| | |
|---|---|
| OFF | Shuts off bleed air from both engines. |
| R ENG | Right engine supplies bleed air. |
| L ENG | Left engine supplies bleed air. |
| BOTH | Bleed air supplied from both engines. |
| AUG PULL | APU augments engine bleed air for ECS operation. Augment works only on the ground and turns off if electrical power is lost or both throttles are advanced to MIL power or greater. The control must be in position L ENG, R ENG, or BOTH for use. |

### Bleed Air Leak Detection

A dual dc powered bleed air leak detection system senses a bleed air leak, displays cautions, and automatically closes the appropriate valves. When the leak detectors cool, the cautions rescind and the valves remain closed.

The fire detection/extinguishing system tests bleed air leak detection (see Fire Warning and Extinguishing, this section). The test closes the primary pressure regulator valves and turns on the bleed air leak cautions, L(R) BLD LEAK and advisories, L(R) BLEED. After the test switch is released the cautions go out, the primary pressure regulator valves remain closed, and the advisories remain on. When the bleed air control is rotated through OFF to BOTH, the primary pressure regulator valves reopen and the advisories go out.

The bleed air leak detection system has three sections. The first is from the primary pressure regulator valve to the secondary pressure regulator valve. If a leak is detected the caution L BLD LEAK or R BLD LEAK appears depending on the side. The primary pressure regulator valve automatically closes. The second is from the secondary pressure regulator valve to the primary heat exchanger and cautions L BLD LEAK and R BLD LEAK both appear if there is a leak. The secondary pressure regulator valve and both primary pressure regulator valves close. The third section is from the air conditioning pack to the secondary heat exchanger. A leak turns on both cautions L BLD LEAK and R BLD LEAK and closes the air conditioning pack flow control valve. The primary and secondary pressure regulator valves remain open and the advisories remain off.

## CONDITIONED AIR

Conditioned air provides cockpit air conditioning, pressurization, defog, and avionics cooling. Warm bleed air is cooled and dried in the air conditioning pack and ducted to the cockpit and avionics bays. The cockpit flow control valve closes and cockpit airflow stops when the canopy is open.

### Cockpit Air Conditioning

Warm bleed air mixes with conditioned air to maintain temperature. Conditioned air enters the cockpit through louvered inlets in the instrument panel, fixed inlets in each leg well, and the defog ducts.

### Cockpit Pressurization

Cockpit pressurization is automatic. The cockpit altitude is ambient pressure to 8,000 feet MSL. Cockpit altitude remains at 8,000 feet to 23,000 feet MSL aircraft altitude. Above 23,000 feet, cockpit pressurization remains at 5.0 psi differential (figure 1-34).

# ENVIRONMENTAL CONTROL SYSTEM

## CONDITIONED AIR

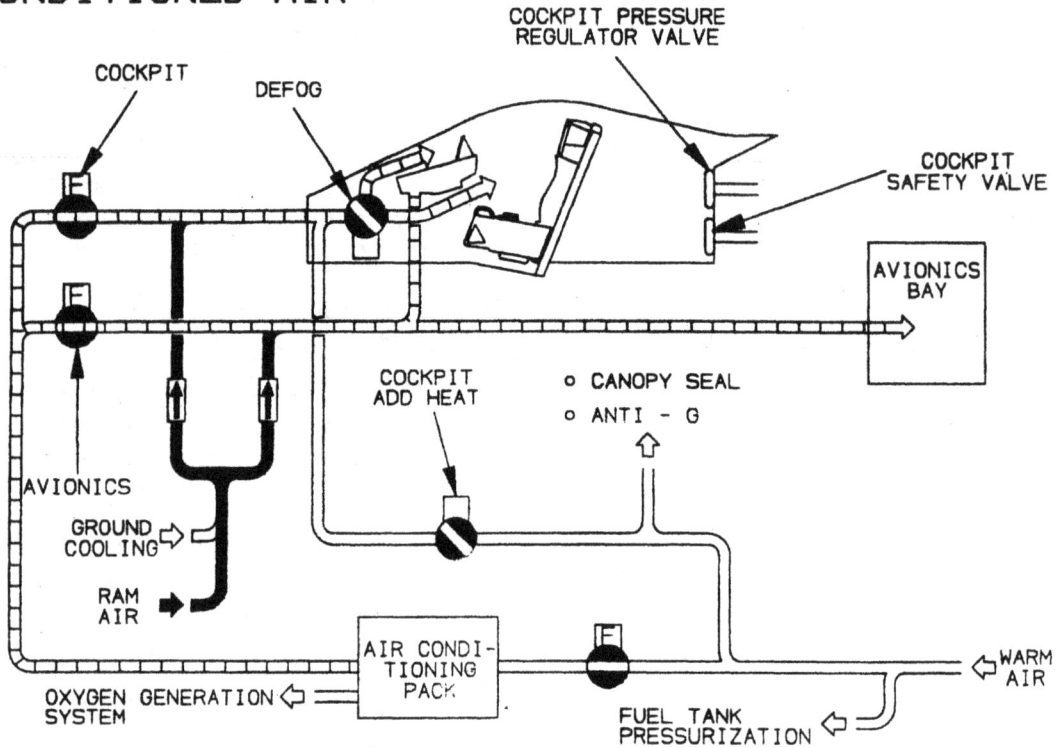

## BLEED AIR

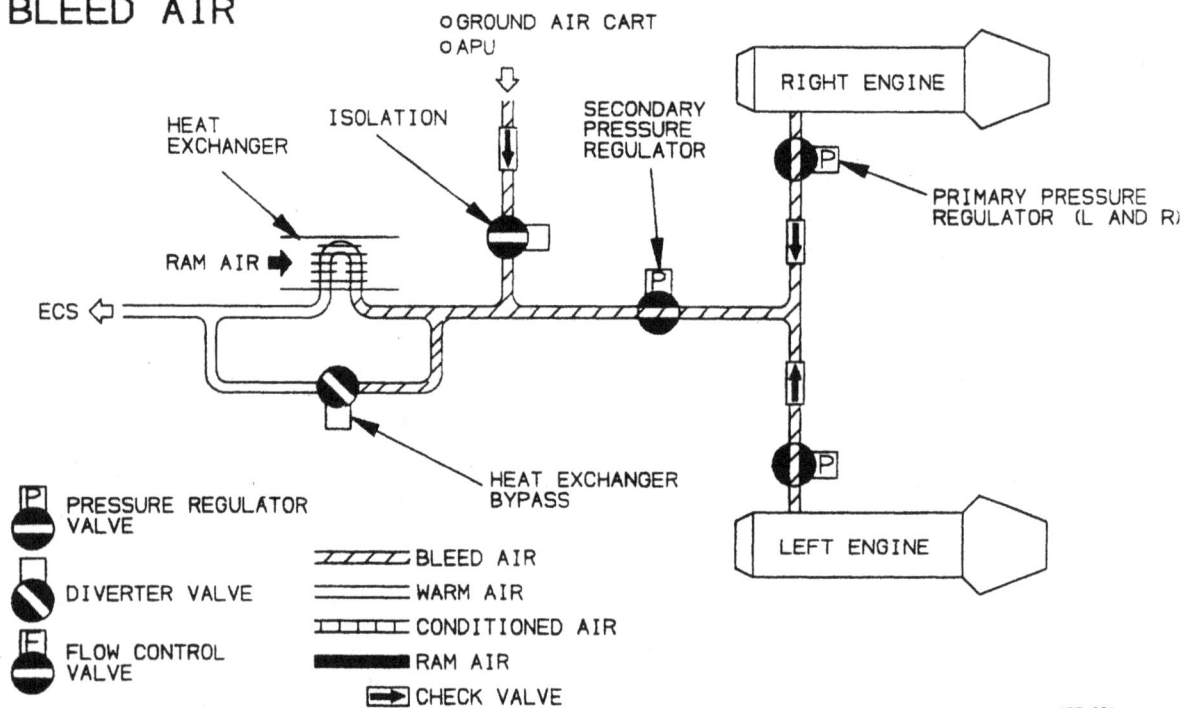

*Figure 1-32*

# ENVIRONMENTAL CONTROL SYSTEM CONTROLS AND INDICATOR

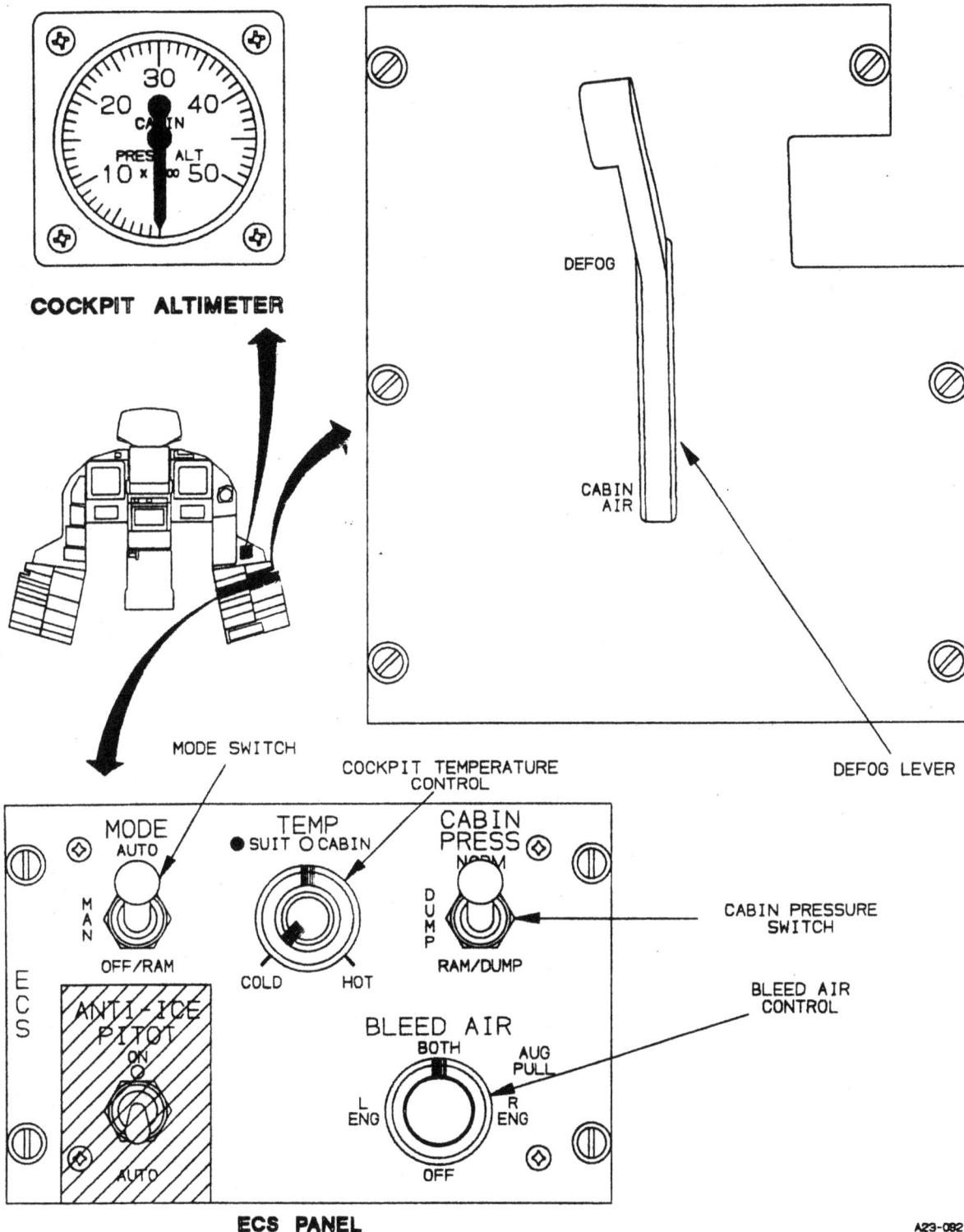

COCKPIT ALTIMETER

DEFOG

CABIN
AIR

DEFOG LEVER

MODE SWITCH

COCKPIT TEMPERATURE
CONTROL

CABIN PRESSURE
SWITCH

BLEED AIR
CONTROL

**ECS PANEL**

A23-082

*Figure 1-33*

## PRESSURIZATION SCHEDULE

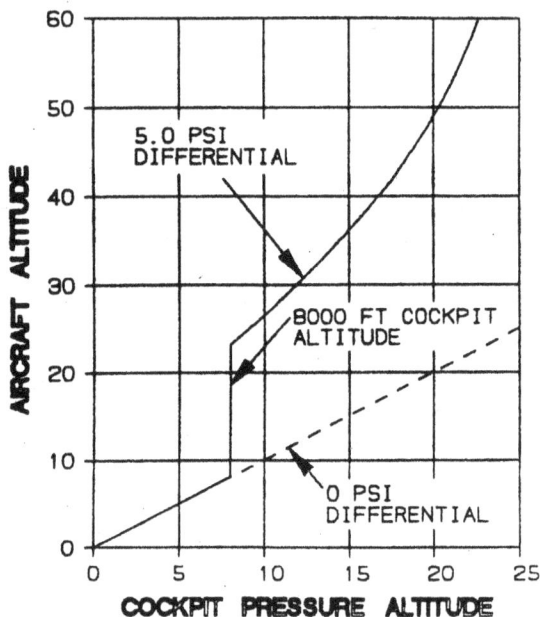

*Figure 1-34*

### COCKPIT ALTIMETER

A cabin pressure altimeter indicates cockpit pressure altitude. The altimeter is marked in 1000 foot increments from 0 to 50,000 feet.

### Defog System

A portion of the conditioned air is diverted to defog ducts along the lower windscreen. The defog lever adjusts the separation of defog air and cockpit air. Defog temperature is the same as cockpit temperature until the lever is within 10 percent of full defog then defog air temperature increases 20 degrees F above selected cockpit air temperature. The defog should be on prior to descent to prevent canopy fogging.

### ECS Controls

The ECS controls are on the ECS panel.

### MODE SWITCH

The three position toggle switch labeled MODE controls the ECS.

| | |
|---|---|
| AUTO | ECS automatically maintains the selected temperature. |
| MAN | Temperature is manually maintained with the temperature control. The ECS pro- |

vides maximum airflow to the cockpit and avionic bays.

| | |
|---|---|
| OFF/RAM | ECS is off. Cockpit pressure regulator valve remains closed and cockpit eventually depressurizes. The emergency cockpit/avionics ram air scoop opens. Oxygen generation is off. |

### TEMPERATURE CONTROL

The control labeled TEMP adjusts cockpit temperature automatically or manually with the position of the mode switch. Turning the control clockwise from COLD to HOT increases the cockpit temperature from 40 degrees F to 110 degrees F. The inner (raised) knob has no function.

### CABIN PRESSURE SWITCH

The three position toggle switch labeled CABIN PRESS controls cockpit pressurization.

| | |
|---|---|
| NORM | The ECS automatically maintains cockpit pressurization. |
| DUMP | Cockpit pressurization is dumped. Airflow to the cockpit and avionics bays continues. Oxygen generation remains on. |
| RAM/DUMP | Conditioned airflow stops. Cockpit pressure regulator valve opens, cockpit pressurization is dumped, and emergency cockpit/avionics ram air scoop opens. Manual cockpit temperature control is available. Oxygen generation is off. |

### DEFOG LEVER

The defog lever moves forward and aft to control the ratio of air used for defog.

| | |
|---|---|
| DEFOG | Maximum conditioned air diverted to defog ducts. Defog supply air temperature increases. |
| CABIN AIR | Maximum conditioned air is directed to cockpit inlets. |

## AVIONICS COOLING

The ECS provides cooling air to the avionics bays. A transient (up to 3 minutes) caution ECS AIR HOT could occur in hot weather following a change from ground cooling at idle power setting to conditioned air cooling at high power setting or with the APU operating in the bleed air augmentation mode. Avionics cooling is automatically maintained between 0 degrees F and 40 degrees F.

## ECS CAUTIONS

ECS cautions appear on the MPCD in amber and turn on the master caution light.

ECS AIR HOT — Avionics cooling is inadequate.

RAM AIR — Emergency ram air scoop is open.

## BLEED AIR CAUTIONS

Bleed air cautions appear on the MPCD in amber and turn on the master caution light.

L(R) BLD LEAK — A bleed air system leak.

## BLEED AIR ADVISORY

The bleed air advisory appears on the MPCD in green.

L(R) /BLEED — Primary pressure regulator valve is closed.

# VEHICLE MANAGEMENT SYSTEM

The vehicle management system (VMS) is a quadruple redundant digital system providing flight control and stability augmentation. Conventional flight control inputs are transmitted electrically to hydraulic control surface actuators. Control surface movements are computer commanded to provide superior handling qualities in a relaxed static stability aircraft.

## CONTROL STICK

The control stick incorporates a trim button, weapon release button, trigger, nosewheel steering button, paddle switch, air refueling release switch, and display selector (figure 1-35).

## RUDDER PEDALS

The adjustable rudder pedals provide yaw inputs, nosewheel steering, and braking. The rudder pedal adjust lever is located at the bottom center of the instrument panel. The adjust handle releases both rudder pedals permitting fore and aft position adjustment.

## TRIM

The control stick trim button controls pitch and roll trim. The yaw trim switch is on the left console. Trim inputs remove control forces without repositioning the stick or rudder pedals.

## CONTROL SURFACES

Control surfaces include leading edge flaps, inboard and outboard trailing edge flaperons, and two movable tail surfaces. The leading edge flaps have one hydraulic actuator each. All other control surfaces have two actuators. Each actuator has a primary and backup source of hydraulic pressure. See Hydraulic System, this section.

### Longitudinal Control

Deflection of both tail surfaces in the same vertical direction with reference to the aircraft (e.g. leading edge down produces aircraft pitch up). At low speed, the control stick commands AOA. At high speed, the pitch axis is a G-command system. Full aft stick provides limit AOA or load factor depending on center of gravity and gross weight. At light gross weights an angle of attack command limiter is employed. A high stick force gradient breakout allows exceeding limit load factor in an emergency.

## *CONTROL STICK*

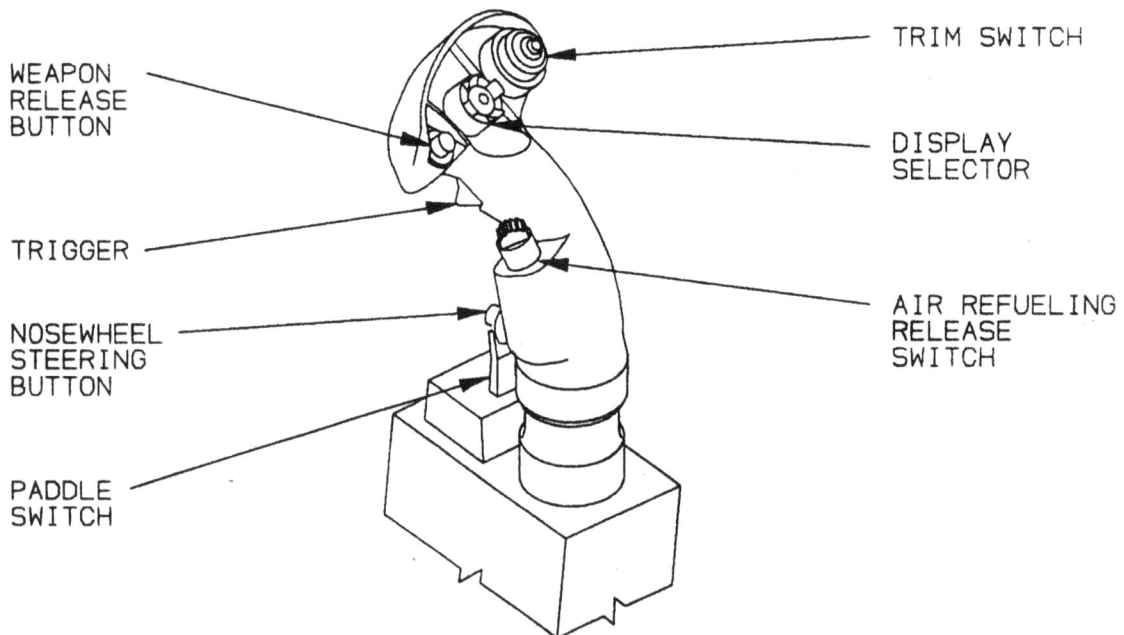

```
WEAPON
RELEASE
BUTTON

TRIGGER

NOSEWHEEL
STEERING
BUTTON

PADDLE
SWITCH

TRIM SWITCH

DISPLAY
SELECTOR

AIR REFUELING
RELEASE
SWITCH
```

A23-098

*Figure 1-35*

## Lateral Control

Differential deflection of flaperons provides bank control. Corresponding tail surface commands balance yaw from the flaperons. At low dynamic pressures, the roll rate limit varies with dynamic pressure and AOA. At high dynamic pressures, roll rate limit varies with load factor.

## Yaw Control

Deflection of both tail surfaces in the same horizontal direction (e.g. leading edge right) produces aircraft yaw (left). Flaperons compensate for roll caused by the tail surfaces. Maximum available yaw command is 10 degrees at low subsonic airspeeds, decreasing to 2 degrees at supersonic airspeeds. Yaw limiting prevents engine inlet flow distortion. Unless yaw is commanded, flight control laws maintain zero sideslip.

## FLAPS

Inboard and outboard flaperons and leading edge flaps extend automatically as required for takeoff, landing, and maneuvering. Flap position varies with Mach and AOA.

The flap emergency switch commands flaps full up or full down.

## SPEEDBRAKES

Differential deflection of inboard and outboard trailing edge flaperons provides high drag flaps (speedbrakes). Above 3.0 G, below 200 KCAS, or above 525 KCAS speedbrakes are unavailable. If speedbrakes were extended when any of these conditions is met, the speedbrakes retract. With the landing gear extended, trailing edge flaperons provide lateral control and flaps. Speedbrakes are unavailable.

The three position speedbrake switch on the inboard throttle should be checked in the forward position for takeoff. The speedbrakes extend after takeoff when passing through 200 KCAS if the speedbrake switch is aft. The switch remains in the selected position.

| | |
|---|---|
| FORWARD | Retracts speedbrakes. |
| CENTER | Stops speedbrakes in any intermediate position. |
| AFT | Extends speedbrakes. |

# VMS COMPUTERS

Four redundant VMS computers receive electrical power from VMS dc buses 1 and 2. Two computers are connected to each bus. One transformer rectifier for each bus converts essential ac power to dc power. If a VMS transformer rectifier fails, the essential dc bus powers the VMS bus. If essential ac and dc power is lost in flight, VMS batteries 1 and 2 provide dc power to the respective VMS bus for approximately 2 minutes. VMS 1 and 2 battery switches on the electrical system control panel connect the VMS batteries.

The VMS computers control:

- Flight control movement,
- Engine operation,
- Boundary layer control door position,
- Nosewheel steering,
- Antiskid,
- Air data,
- Flight control related flight test data.

VMS computers receive hardwired inputs from the control stick, rudder pedals, throttles, anti-skid panel, brake controls, and sensors for weight on wheels, gear position, and gear handle position. The computers transmit commands directly for flight control movement, engine boundary layer control door positioning, and nosewheel steering. The computers communicate via a VMS data bus with the air data computers, engine control computers, flight control gyros and accelerometers, brakes, and standby flight display. The computers communicate via the avionics bus with the fuel system and avionics displays. The computers communicate via the avionics bus with the armament control system to prohibit door operation or armament firing with hydraulic malfunctions.

Variable gains within the VMS computers limit control deflections depending on airspeed. If total air data failure occurs, the vehicle management computers use the last valid air data. The fixed gain switch provides fixed gains for operation above 350 KCAS or for operation between 350 KCAS and 200 KCAS. A third set of fixed gains for landing is also available. For recovery from out of control situations, the pilot has full control surface deflection capability when the flight control override switch is engaged and the paddle switch is pressed and held.

## AIR DATA

The air data system consists of four sets of static ports, one total pressure port, two Pitot tubes, and air data computers. All air data ports are flush mounted and quadruple redundant.

### AOA Static Ports

AOA static ports are located top and bottom on centerline in the radome. Differential pressure provides angle of attack.

### Sideslip Static Ports

Sideslip static ports are located in the radome off centerline. Differential pressure provides angle of sideslip. Combined pressure from both sides provides static pressure for airspeed and altitude computation.

### Total Pressure Ports

A set of total pressure ports is in the nose. Total pressure is measured but not used for computation pending flight verification of data.

### Pitot Tubes

Two L-shaped Pitot tubes on the lower fuselage aft of the radome provide total pressure for airspeed computation.

### Air Data Computers

An air data computer is provided for each VMS computer. Each air data port transmits pneumatic pressure to each of the four computers. Computer air data is transferred via the VMS data bus.

An air data validity check occurs on takeoff at approximately 45 knots and compares airspeed, wheelspin, and INS velocity. An out of tolerance condition causes caution FCS 1ST.

### Pitot Static Heat Switch

The Pitot static heat switch is located on the ECS panel (figure 1-36).

| | |
|---|---|
| AUTO | All air data sensors (flush and L head) are ac electrically heated when airborne. |
| ON | Permits ground operation of air data sensor heaters. |

## VMS CONTROLS

The VMS controls are on the left console and the control stick (figure 1-37).

### Flight Control Override Switch

The two-position guarded toggle flight control over-override switch bypasses flight control limits to provide maximum control surface command capability. Raising the guard and placing the switch to the forward position ENGAGE arms flight control override. Pressing the control stick paddle switch engages flight control override. Releasing the paddle switch or returning the flight control override switch to OFF cancels flight control override.

### VMS/Engine Reset Switch

The VMS/engine reset switch is a two-position toggle switch, spring loaded to OFF. The switch should be activated for any VMS or engine caution accompanied by a message RESET VMS. Activating the VMS/engine reset switch extinguishes the legend RESET VMS, on the MPCD.

### Flap Emergency Switch

The flap emergency switch is a three position, lever locked switch.

| | |
|---|---|
| EMERG UP | Commands all flaps to the full up position. |
| NORM | Normal VMS control of flap position. |
| EMERG DN | Commands flaps to the landing position. |

### Fixed Gain Switch

The three position, lever locked fixed gain switch commands VMS computer fixed gains in case of total air data failure. With landing gear extended, either position HI A/S or LO A/S gives landing gains.

| | |
|---|---|
| HI A/S | Commands fixed gains for use above 350 KCAS. |
| OFF | Normal gain scheduling by VMS computer or gains based on last valid air data before total air data failure. |
| LO A/S | Commands fixed gains for use below 350 KCAS. Recommended minimum airspeed is 200 KCAS. |

## PITOT STATIC HEAT SWITCH

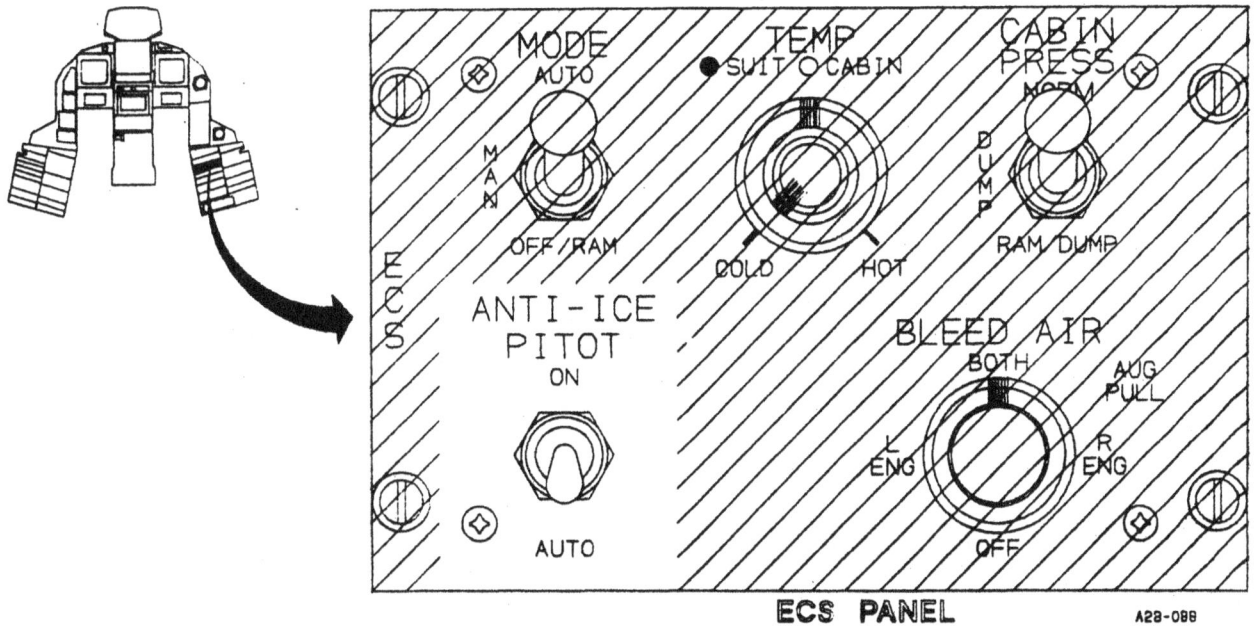

**ECS PANEL**

A28-088

## VEHICLE MANAGEMENT SYSTEM CONTROLS

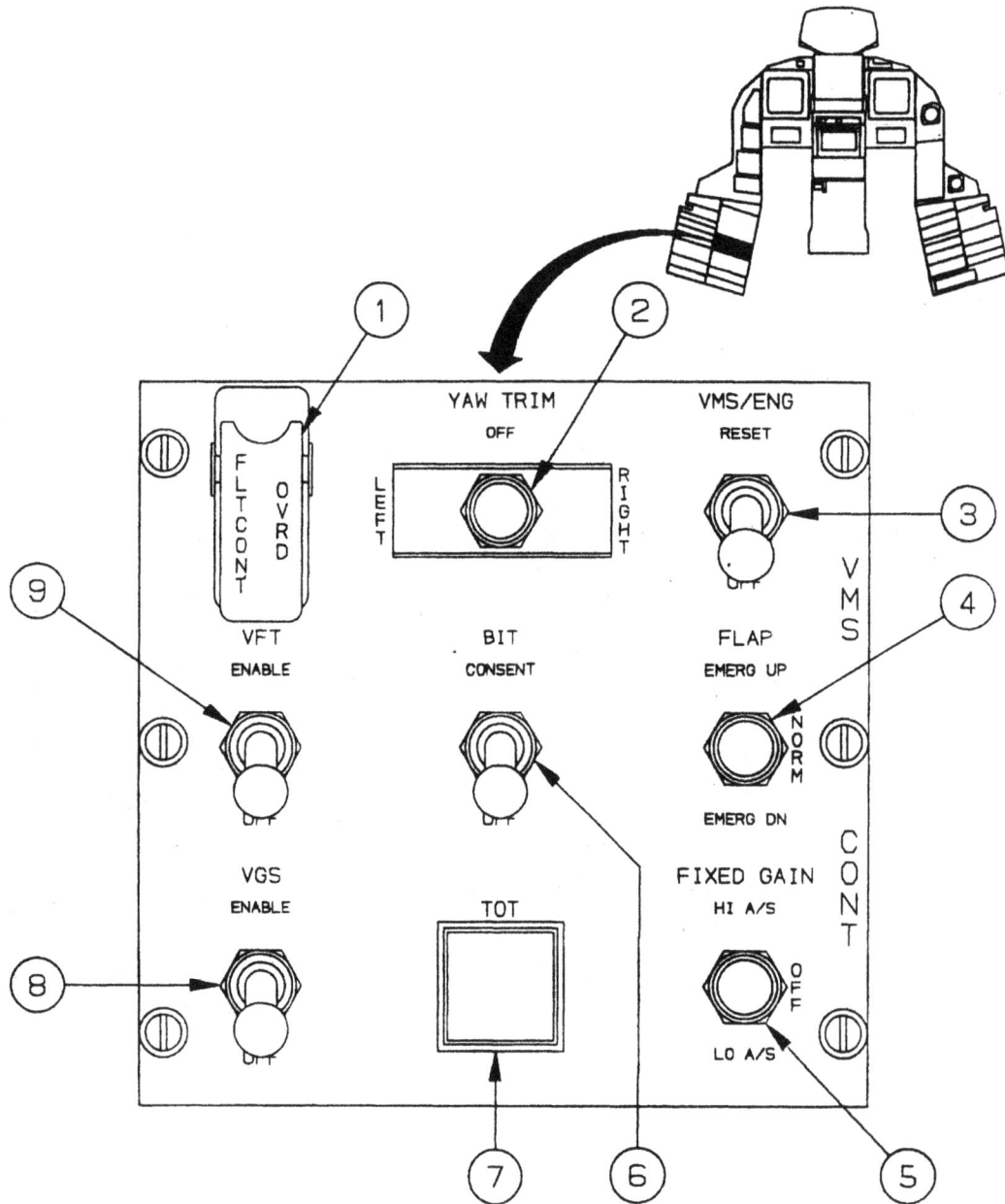

1. FLIGHT CONTROL OVERRRIDE SWITCH
2. YAW TRIM SWITCH
3. VMS/ENGINE RESET SWITCH
4. FLAP EMERGENCY SWITCH
5. FIXED GAIN SWITCH
6. BIT CONSENT SWITCH
7. TAKEOFF TRIM BUTTON
8. VARIABLE GAIN ENABLE SWITCH
9. NOT USED

A23-087

*Figure 1-37*

## BIT Consent Switch

The BIT consent switch is a two-position toggle switch spring loaded to the off position. The switch must be held in the consent position while commanding VMS IBIT via the MPCD.

## Takeoff Trim Button

The takeoff trim button sets trim for takeoff. When all trim are set properly, the integral indicator light comes on for approximately three seconds. Failure of the indicator light to come on means takeoff trim is not set properly.

## Variable Gain Enable Switch

The two-position lever-locked variable gain enable switch allows selection of variable gain sets using the MPCD VMS gain select display when the position ENABLE is selected.

## VMS CAUTIONS

| | |
|---|---|
| AIR DATA MC | Degraded air data. Air data to VMS computers still valid. Further degradation could cause FX GAINS caution. |
| CG | Computed CG out of limits. |
| ENVELOPE | Exceeding: <br> • Airspeed Mach. <br> • Angle of slip. <br> • Angle of attack. <br> • G. |
| FCS 1ST | • Flight control system fault. <br> • RESET VMS caution appears. |
| FCS 2ND | Two flight control system like faults. |
| FCS 3RD | Three flight control system like faults. |
| FLAPS | Flap system malfunction. |
| FX GAINS | • Total air data failure. <br> • Last valid air data used by VMS computers until fixed gain switch activated. |
| RESET VMS | Vehicle management system or engine fault. |
| TAIL ACT | Malfunction of tail actuation system. |
| VMS HOT | VMS computer overheated or engine fuel controller overheated. |

## VMS ADVISORY

| | |
|---|---|
| VMS BIT | VMS BIT failure detected (ground only). |

# CONTROLS AND DISPLAYS

Controls and displays provide for monitoring and control of aircraft systems. The controls and displays system consists of multipurpose color displays (MPCD), head up display (HUD), multipurpose display processor, and upfront control (UFC). Data input is through the UFC. MPCD displays can be selected or manipulated with control stick switches. See system descriptions in this section for specific controls and displays applications.

The multipurpose display processor, central computer and avionics interface units process data and provide redundancy management for controls and displays components. The multipurpose display processor and interface units manage power for controls and displays and other avionics.

# UPFRONT CONTROL

The upfront control (UFC) provides monitoring and control of communications, navigation, and identification, through menus and data displays (figure 1-38). The UFC also supports mission recording and flight test.

The UFC has six lines of information on liquid crystal displays (LCD). Each line has two or three data fields. The center data field on the bottom line is the scratchpad, for data entry. UHF radio information appears on the bottom two lines of all menus and data displays.

Multifunction buttons flank the top five lines of the display area. The buttons provide data entry, control of avionics, and selection of menus and displays.

If data is not available for a system or component, because of avionics or electrical failure, related data fields, menus, or data displays are not accessible.

An asterisk or colon can appear in a data field immediately adjacent to the button. An asterisk indicates a selected option. A colon indicates that the identified component is powered.

Keyboard functions include entering alphanumerics, clearing the display area, and selecting top level menus or data displays. Other functions affect UHF, IFF, and mission recording.

Knobs on the UFC control LCD brightness, UHF volume, and UHF channel selection.

## UPFRONT CONTROL REDUNDANCY

Two avionics interface units (AIU) provide data and power control for the upfront control, with single unit operation supporting most functions. If inter-

interface unit 1 fails, control of UHF 1, KY-58, TACAN, and IFF is lost. UHF 2 control is lost if interface unit 2 fails. UFC dc power is from the right main bus.

See Backup Avionics Operations for the effect of central computer failure on UFC functions.

## KEYBOARD

Alphanumeric keys enter characters into the scratchpad. Pressing the shift key (SHF) permits entry of the upper character on the next key pressed. Simultaneous use of the keys is not necessary.

The keys labeled GREC are for UHF radio.

| | |
|---|---|
| I/P | Selects identification (IDENT) function of IFF. |
| MENU | • Selects menu 1.<br>• Pressing again selects menu 2. |
| DATA | • Selects data 1 display.<br>• Pressing again selects data 2 display. |
| CLR | • Clears last character entered in scratchpad. Pressing twice clears scratchpad.<br>• If scratchpad is flashing, the key must be pressed three times to clear scratchpad.<br>• When scratchpad is blank, clears top four lines of display area. Pressing twice clears display area. |
| MARK | Selects mark point data display. Enters present position and time as next available mark point. |

## UPFRONT CONTROL MULTIFUNCTION BUTTONS

Button functions include data entry, control of avionics, and selection of menus, data displays, and other options. Functions vary according to the options selected. In most cases, a button can either enter data or select another menu or display, depending on scratchpad contents.

## UPFRONT CONTROL

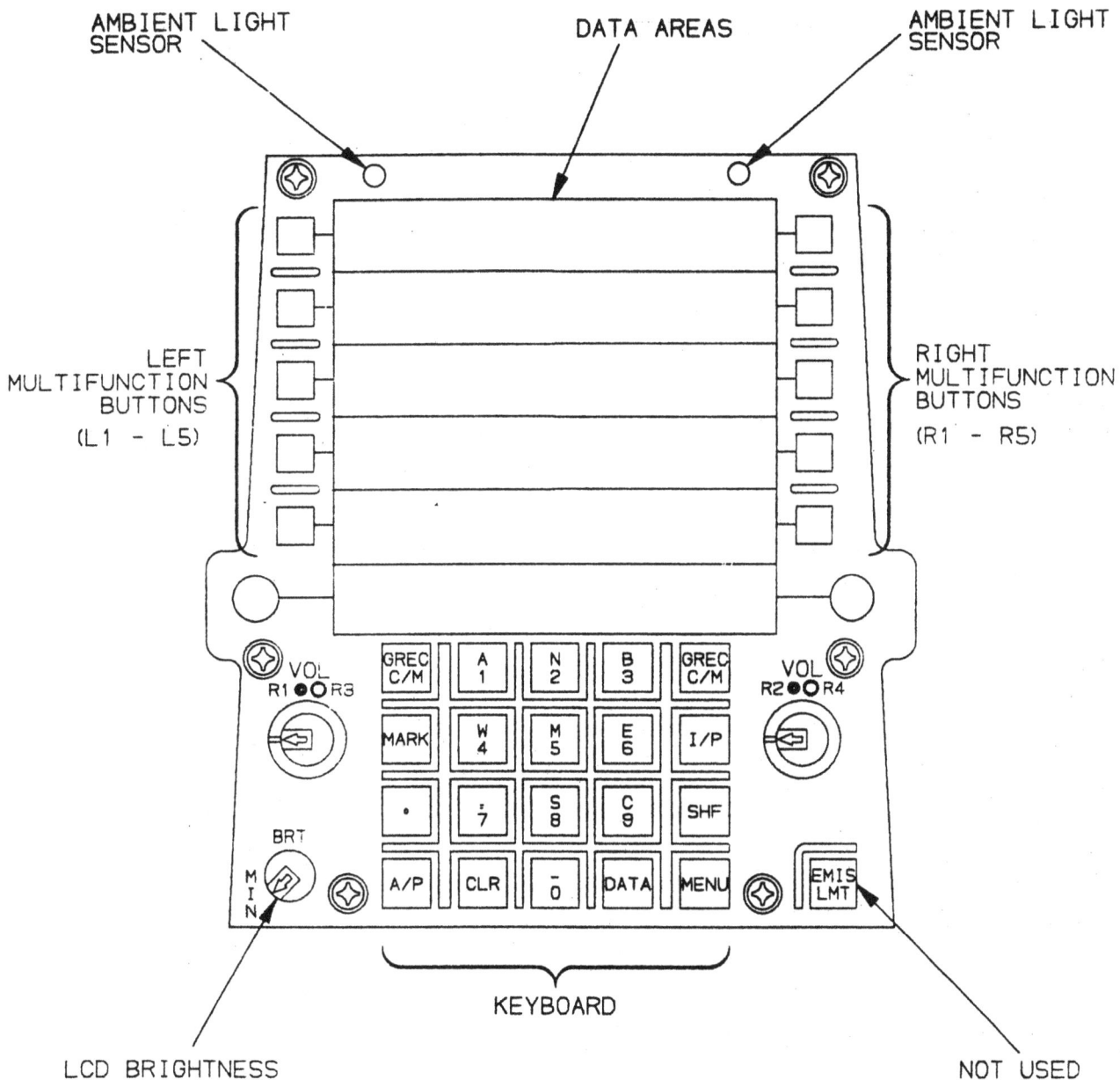

AMBIENT LIGHT SENSOR          DATA AREAS          AMBIENT LIGHT SENSOR

LEFT MULTIFUNCTION BUTTONS (L1 - L5)

RIGHT MULTIFUNCTION BUTTONS (R1 - R5)

| GREC C/M | A 1 | N 2 | B 3 | GREC C/M |
| MARK | W 4 | M 5 | E 6 | I/P |
| . | ; 7 | S 8 | C 9 | SHF |
| A/P | CLR | - 0 | DATA | MENU |

VOL R1 R3

VOL R2 R4

BRT
MIN

EMIS LMT

KEYBOARD

LCD BRIGHTNESS

NOT USED

A23-011

**Figure 1-38**

Descriptions of multifunction buttons include adjacent data fields.

Pressing a multifunction button with data in the scratchpad causes a validity check of the data for entry into the adjacent data field. Valid data is entered for systems use. If the data is invalid, the scratchpad entry flashes. When the scratchpad is blank, a button can select an associated menu or data display.

Entries for time, latitude, longitude, and UHF frequencies have fields requiring fixed numbers of digits. Leading zeroes are required, trailing zeroes are entered automatically. If fields are properly filled, colons or decimals are generally not required. For example, time entered as 1 would appear as 10:00:00. Time of 01:00:00 can be entered as 01.

### LIQUID CRYSTAL DISPLAY BRIGHTNESS CONTROL

The brightness control regulates backlighting for the display area. Sensors disable the control during normal daytime conditions.

### UPFRONT CONTROL BLANKING

When the scratchpad is blank, the clear key clears the top four lines of data. Pressing again clears the bottom two lines.

Either UHF button restores communications information to the bottom lines. Pressing any other multifunction button with the scratchpad blank restores all information to the display area.

With the top four lines blank, all keys and buttons are active, and data entry is possible only for UHF radios. When all lines are blank, no data entry is possible. On the keyboard, the mark, data display, menu, and IDENT keys are active.

## UPFRONT CONTROL DATA DISPLAYS AND MENUS

The upfront control has top level menus and data displays selected from the keyboard, and subordinate menus selected with multifunction buttons. Data displays have fuel and navigation information. Menus provide monitoring and control of flight planning, mission recording, navigation, and systems operation. Figure 1-39 shows the multiple levels of subordinate menus, discussed in appropriate system descriptions, this section. Menus for navigation points are accessible from both menus and data displays. The shows menu and display availability if the multipurpose display processor performs central computer functions.

### NAVIGATION POINTS

Navigation points are stored as sequence points and mark points. Mark points record where designated actions occurred. Sequence points are steer points, targets, aim points, and offset points. Points on the planned route are steer points or targets. Aim and offset points are not used.

Points for three routes, identified A, B, and C, can be stored through the UFC. The point identifier includes the sequence number and route and indicates the type of point. Steer points have the sequence number and the route letter such as 19A. Target identifiers have a decimal point after the number such as 20.A.

Data 1 display and menu 1 show the currently selected route point (STR). Navigation systems provide guidance data for the current steer point.

### DATA 1 DISPLAY

The data 1 display provides for selection and presentation of time, altitude, and navigation information. See Navigation Systems, this section.

### MENU 1

Menus permit monitoring and control of systems. Access to subordinate menus is through menus 1 and 2.

| | |
|---|---|
| Bingo Fuel | Shows bingo fuel in pounds, for activation of ***BINGO FUEL*** voice warning. |
| | • Enters scratchpad value from 0 to 18000. |
| Steer Point | Shows current steer point. |
| | • Selects new steer point with valid scratchpad entry. |
| | • Selects steer point menu for selected point. |
| TACAN | Shows current TACAN channel. |
| | • Enters valid channel number from scratchpad. |
| | • Selects TACAN menu. See Radio Navigation, this section. |

## UFC DISPLAYS AND MENUS

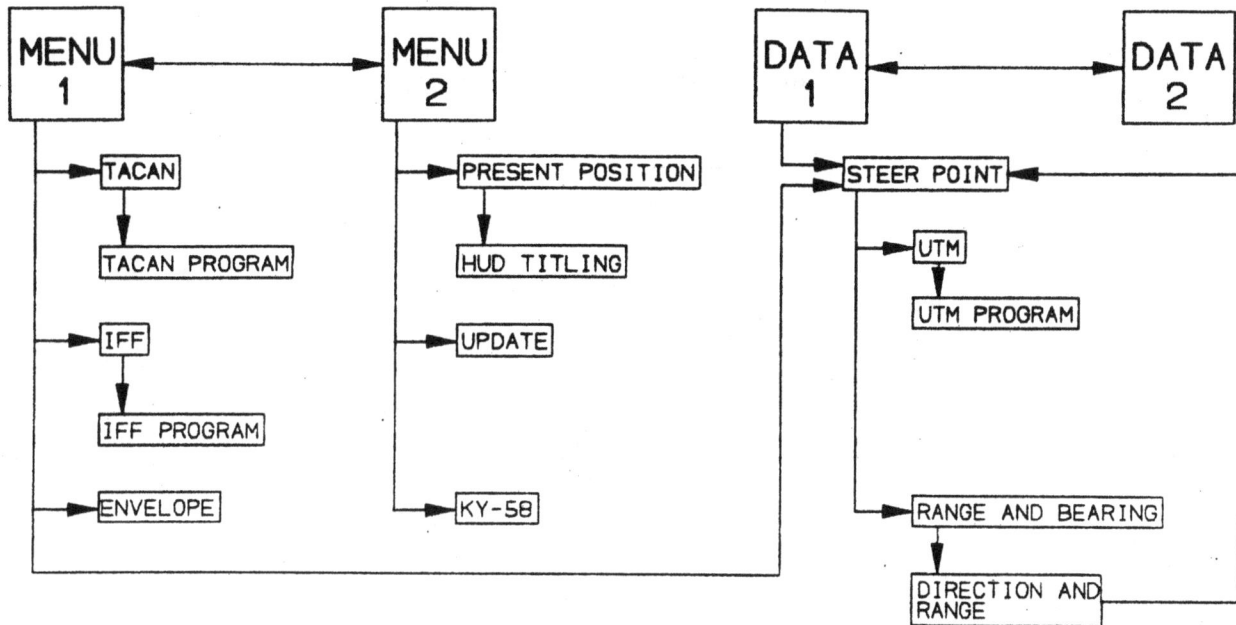

NORMAL OPERATION

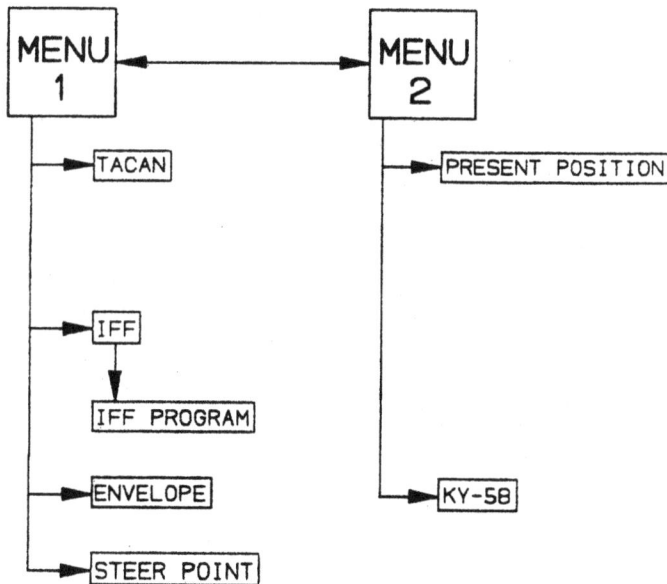

BACKUP OPERATION

A23-021

*Figure 1-39*

IFF — Shows selected IFF modes.
- Enters valid scratchpad mode setting.
- Selects IFF menu. See Communication Systems, this section.

Barometric Pressure — Shows barometric pressure setting for pressure altitude indications.

Enters scratchpad value for 28.10 to 31.00 inches of mercury. The decimal point is necessary.

Envelope — Selects envelope menu.

## MENU 2

Menu 2 has head up display and communications functions, and gives access to communications and navigation menus.

HUD Reticle —
- Shows HUD reticle depression in mils.
- Enters valid scratchpad value as HUD reticle depression.

Present Position Source —
- Shows present position source.
- Selects present position menu.

KY-58 — Selects KY-58 menu.

Update — Selects update menu.

Transmitter Select — Toggles between single or dual radio transmission. Asterisk indicates dual transmission.

## ENVELOPE MENU

Flight parameters set with the envelope menu (figure 1-40) affect the caution ENVELOPE. Exceeding a set parameter, except weight, triggers the caution and associated tone.

CAS — Sets maximum calibrated airspeed in knots. Valid entries are 0 to 999.

Mach — Sets maximum Mach number, from 0.00 to 3.00.

AOA — Sets angle of attack in degrees, from -20.0 to 90.0.

AOSS — Sets angle of sideslip, from -32.0 to +32.0 degrees.

Aft CG — Sets percent MAC for maximum aft CG, from 32.0 to 43.0.

G — Sets G, from -5.0 to 13.5.

CG — Sets zero-fuel center of gravity in inches. Valid entries are 560.0 to 600.0.

Weight — Sets aircraft zero-fuel weight in pounds. Valid entries are 35000 to 50000. Does not affect the caution.

With the aircraft on the ground (weight-on-wheels), weight and CG data are entered for VMS use by simultaneously pressing the VMS BIT consent switch and squeezing the control stick trigger to the second detent.

## ENVELOPE MENU

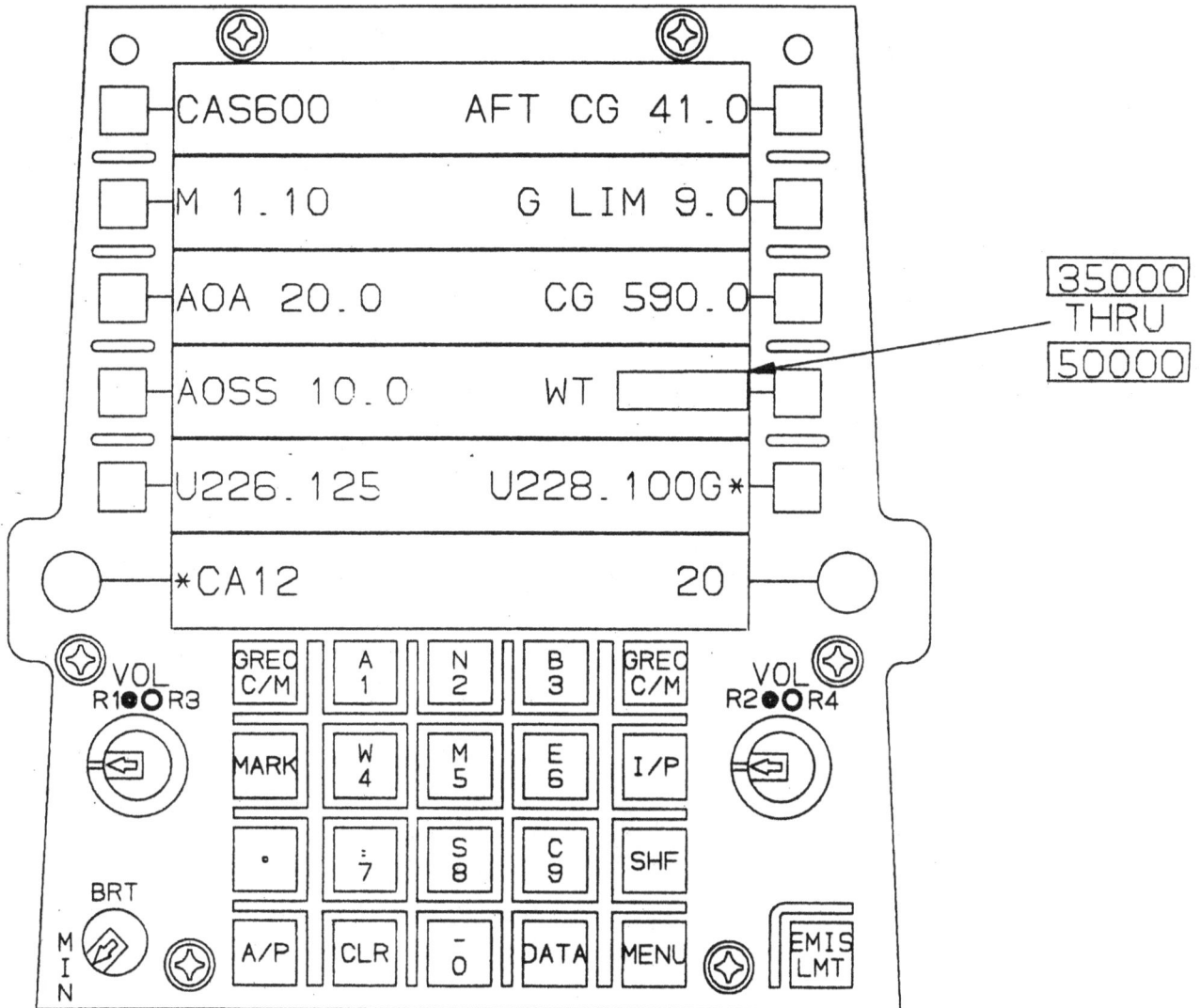

35000
THRU
50000

A23-015

# HEAD UP DISPLAY

The head up display (HUD) provides electro-optical flight conditions and attack geometry. Depending on mode operation this information is used for flight control, navigation, and aircraft handling evaluation. There are two modes of operation navigation and flight test. See figure 1-41 for typical HUD display, symbols, and controls.

A wide field-of-view combining glass (18 degrees in elevation, 28 degrees in azimuth) presents flight parameters and weapons sight and symbology. The color of the projected imagery is green.

Magnetic heading, airspeed, altitude, flight path vector, load factor (g), pitch scale, aircraft symbol, and angle of attack information is shown. Information is controlled by means of the HUD control panel, up front control panel, and armament control switch.

## HUD CONTROL PANEL

Adjustment for the HUD is located on the HUD control panel. Master modes labeled A/A, A/G, NAV, and INST are inoperative in the PAV. Video controls, brightness and contrast, are operative but non-functional.

### Brightness Control

Rotating the knob labeled BRT clockwise from the OFF position turns on setting OFF removes power from the HUD.

### Symbols Control

The three position switch labeled SYM restores/removes symbology or selects flight test HUD display.

| NORM | Restores symbology. |
| REJ 1 | Declutter function. Removes heading scale, pitch ladder, and bank scale from HUD. |
| REJ 2 | Selects flight test HUD. |

### Intensity Control Switch

| DAY | Bright intensity. Daytime use. |
| AUTO | Adjusts brightness levels according to light intensity. |
| NIGHT | Reduces maximum intensity. |

### BIT Indicator

The BIT indicator is a magnetically controlled ball that rolls over to show white if the HUD has failed.

In normal operation the ball is black. BIT ball holds last position when power is removed. Initiating built-in-test resets the indicator. The unlabeled button above the BIT indicator is a maintenance test switch.

## HUD SYMBOLS

Symbology presentation is dependent on program software and is expected to vary during the test program. All symbols may not appear has shown in the example displays.

## HUD FORMATS

HUD formats are for navigation and a modified format for flight test, and variable with landing gear up or down.

### Navigation HUD format

The navigation mode is selected when the brightness control is out of off and the reject switch is in normal or REJ 1. Navigation mode steering and display symbology is controlled by the navigation system selected on the electronic HSI.

The word OFF appears as a cue for invalid airspeed, altitude, or navigation information. Mach number and angle of attack do not appear if information is invalid.

### Flight Test HUD format

The flight test HUD format is selected by moving the reject switch to the REJ 2 position. The flight test display is similar to the navigation format with the addition of a sliding acceleration (g) scale and digital sideslip information added. An index marker can be set on the acceleration scale by using the command heading pushbuttons on the HSI (use a multiplier of 10, e.g., 10 degrees = 1g), prior to selecting the flight test format. If REJ 2 is deselected and a new command heading is entered, the command g is also changed when REJ 2 is reselected.

When the gear is down, the acceleration scale is replaced by a vertical velocity scale and digital vertical velocity information is presented under the altitude window. The vertical velocity index marker can be set in the same manner as the g index marker on the gear up flight test display.

## GUN RETICLE

The gun reticle is used for evaluation of aircraft handling and maneuverability. With the programmable armament control system installed, the weapon select switch on the throttle in the aft gun position causes the reticle to appear on the HUD. With the weapons select switch in the center SRM or forward MRM

## TYPICAL HUD DISPLAY AND CONTROL PANEL

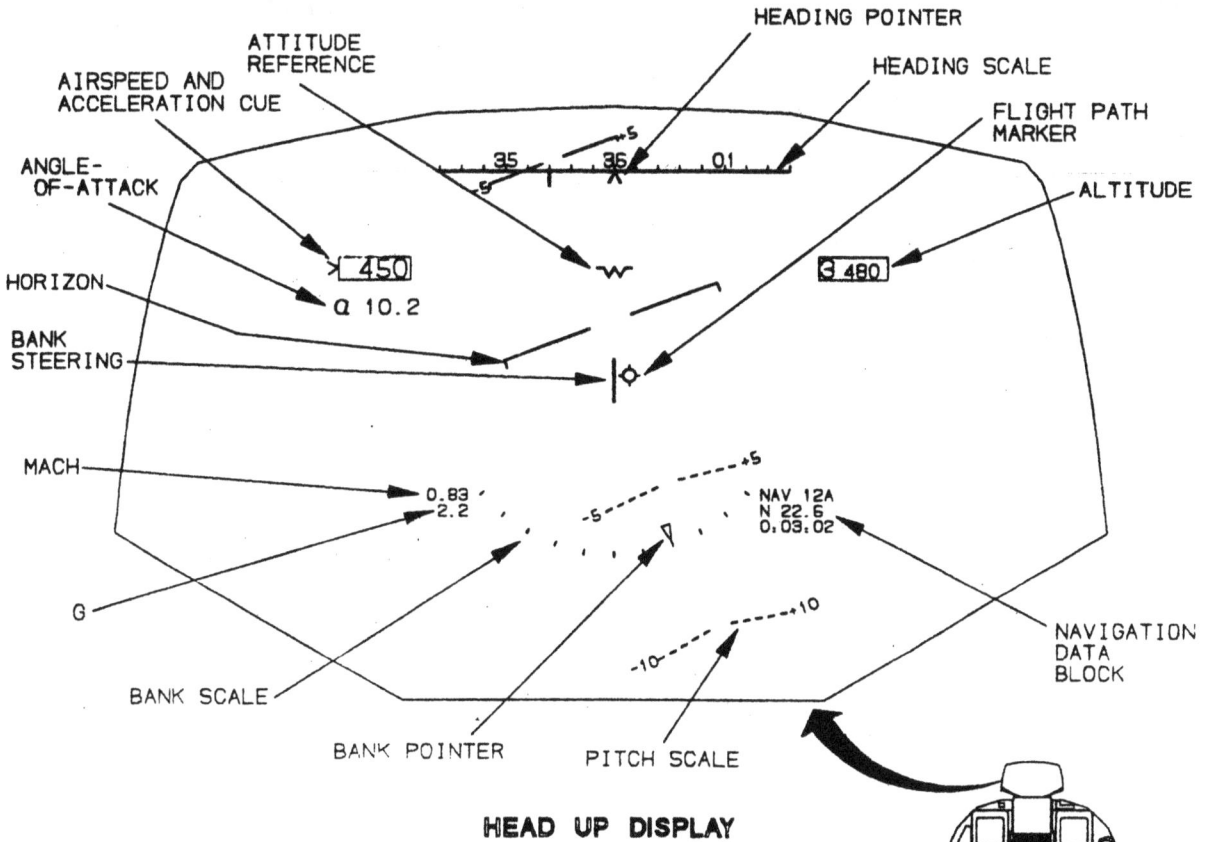

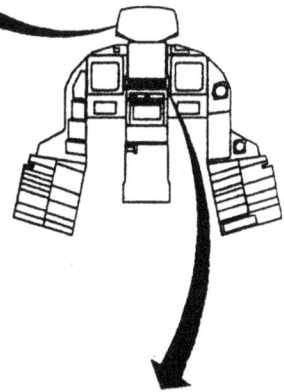

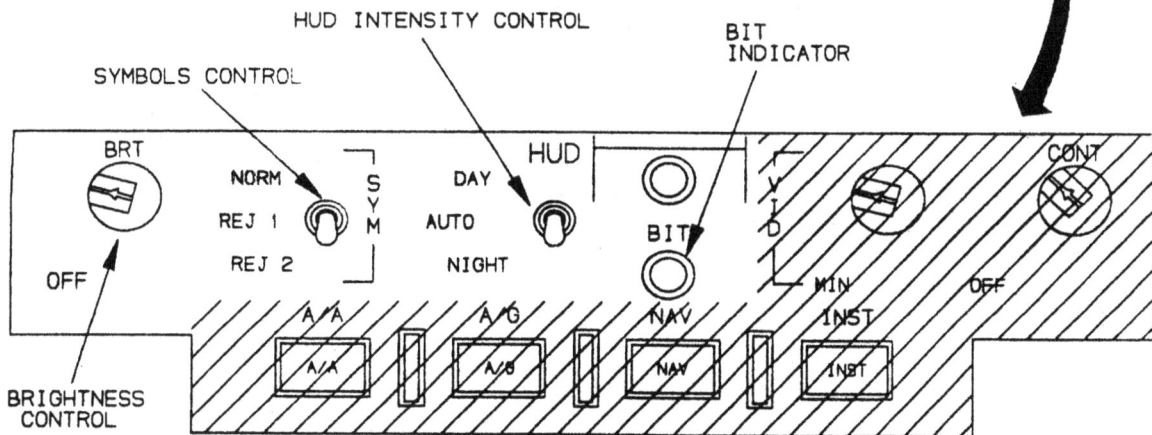

HEADING POINTER

HEADING SCALE

FLIGHT PATH
MARKER

ATTITUDE
REFERENCE

AIRSPEED AND
ACCELERATION CUE

ALTITUDE

ANGLE-
OF-ATTACK

450

3 480

HORIZON

α 10.2

BANK
STEERING

◇

MACH

0.83
2.2

+5

-5

NAV 12A
N 22.6
0, 03, 02

G

-5

BANK SCALE

-10

-10

NAVIGATION
DATA
BLOCK

BANK POINTER

PITCH SCALE

**HEAD UP DISPLAY**

HUD INTENSITY CONTROL

BIT
INDICATOR

SYMBOLS CONTROL

BRT

NORM

S
Y
M

HUD

DAY

CONT

REJ 1

AUTO

BIT

OFF

REJ 2

NIGHT

MIN

OFF

INST

A/A

A/G

NAV

A/A

A/G

NAV

INST

BRIGHTNESS
CONTROL

**HUD CONTROLS**

A29-016

*Figure 1-41*

position, the gun reticle is deselected. If the programmable armament control system is not installed, the gun reticle appears regardless of weapons select switch position. To depress the reticle out of the HUD main field of view, use the HUD RET option on the UFC Menu 2.

# MULTIPURPOSE COLOR DISPLAY

Two multipurpose color displays (MPCD) provide essential flight and navigation information and permit monitoring and control of systems with a variety of displays on cathode ray tubes.

Primary flight information displays are the electronic attitude director indicator (ADI), electronic horizontal situation indicator (HSI), and tactical situation display (TSD). Displays are available for engine, armament, fuel, and vehicle management systems. Built-in test is available through the MPCD for avionics, fuel, VMS, and armament systems.

Video tape recording of color and head up displays can be selected from MPCD. A HUD repeater display is available, with an option to view the onboard camera picture.

Any display can appear on either MPCD. A program function permits selecting three displays for each MPCD with the display selector on the stick grip.

Caution and advisory messages appear on one MPCD, usually the right. Messages can be selectively moved to the other MPCD by pressing the master caution light and simultaneously moving the display selector toward the desired side.

Each MPCD (figure 1-42) consists of a cathode ray tube, 20 pushbuttons, a display power control, contrast and brightness controls, and a BIT indicator. The contrast control normally is not used. The figure shows the MPCD menu.

MPCD turn-on is automatic or manual, with initial brightness set to an intermediate level. The screen retains the last display selected. The left and right MPCD receive dc power from the corresponding main dc bus.

## MPCD DISPLAY SCREEN

The cathode ray tube display area is five inches square. Alphanumerics and symbols are normally white, orange, blue, or green. Cautions and indications requiring attention are yellow. Except on the armament display, red legends indicate input information is invalid. Invalid information does not appear.

## MPCD CONTROLS

Power and brightness controls are rocker switches, spring-loaded to neutral center. The brightness control has arrows for increase (up) and decrease. The power switch is effective after initial automatic powerup. If an MPCD is turned off with the power switch, the switch must be used to restore power.

## MPCD BIT INDICATOR

The built-in test indicator normally is black, showing white if the MPCD fails BIT. Successful initiated BIT resets the indicator. The last BIT indication remains when the MPCD is turned off.

## MPCD BUTTONS

Buttons around the edge of the display area have associated fields that can contain a label and symbols or other data. Labels are in fields immediately adjacent to buttons. A labeled button selects the indicated display, manipulates information on the current display, or controls system operations. The menu display, selected with the menu button from any other display, permits selection of displays.

## MPCD MENU

The MPCD menu controls selection of other displays. Pressing a labeled button selects the indicated display or establishes a sequence for selecting displays with the control stick display selector.

## MPCD DISPLAY SEQUENCING

Pressing the program button boxes the label PROG and starts the sequencing program for that MPCD. Pressing buttons for other displays sets the sequence for calling the displays. A boxed number next to the label indicates the sequence number, from 1 to 3. A sequence can be changed by pressing the button for a selected display to erase the boxed number, and selecting another display. The number reappears by the newly selected label.

Pressing the boxed program button saves the sequence, erases the box from the label, and permits normal selection of another display. Moving the control stick display selector left or right toward an MPCD steps displays on that side through the programmed cycle.

Movement of the display selector to the right places the ADI on the right color display. Programmed sequencing is then available by moving the selector to the right at intervals of not more than a few seconds.

## ELECTRONIC ADI

The electronic ADI (figure 1-43 and 1-44) provides attitude, heading, altitude, vertical velocity, turn and slip, angle of attack (AOA), air or ground speed, and bank steering data. Most symbols and characters are white when data is valid, turning red if data is invalid or missing.

## MULTIPURPOSE COLOR DISPLAY

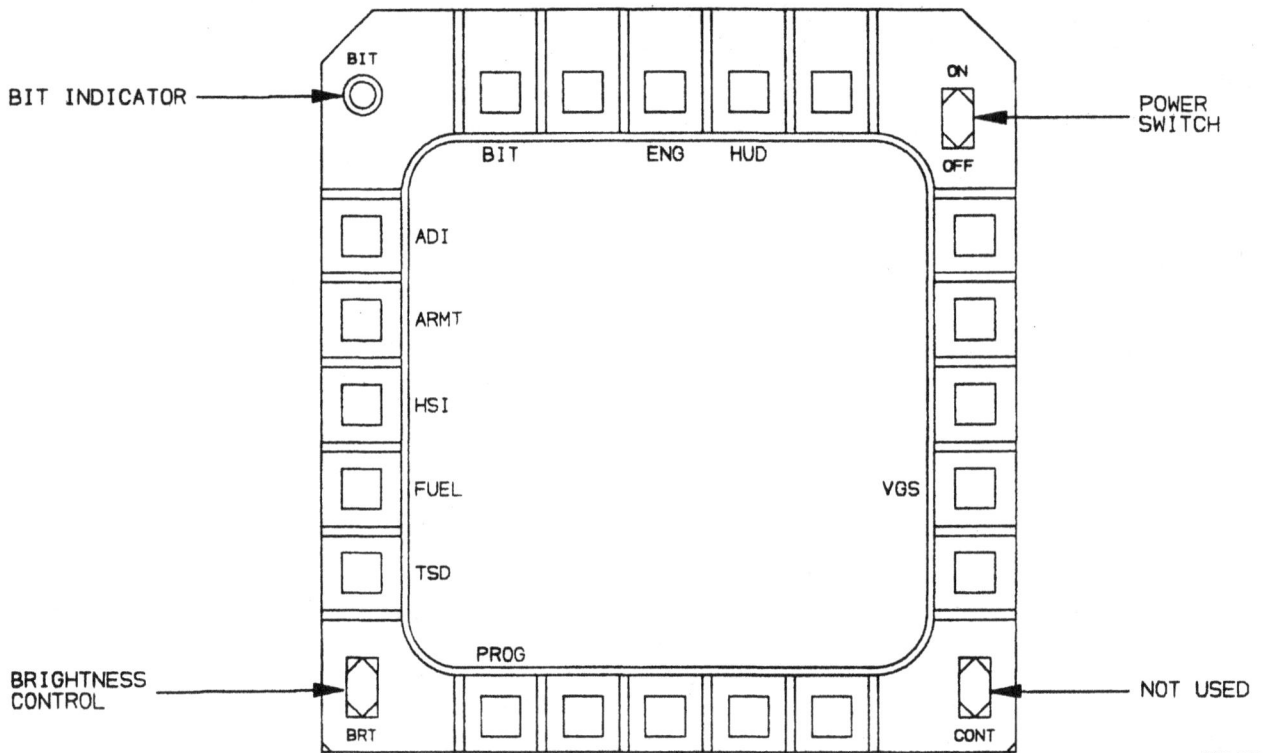

Figure 1-42

A23-012

## ELECTRONIC ADI

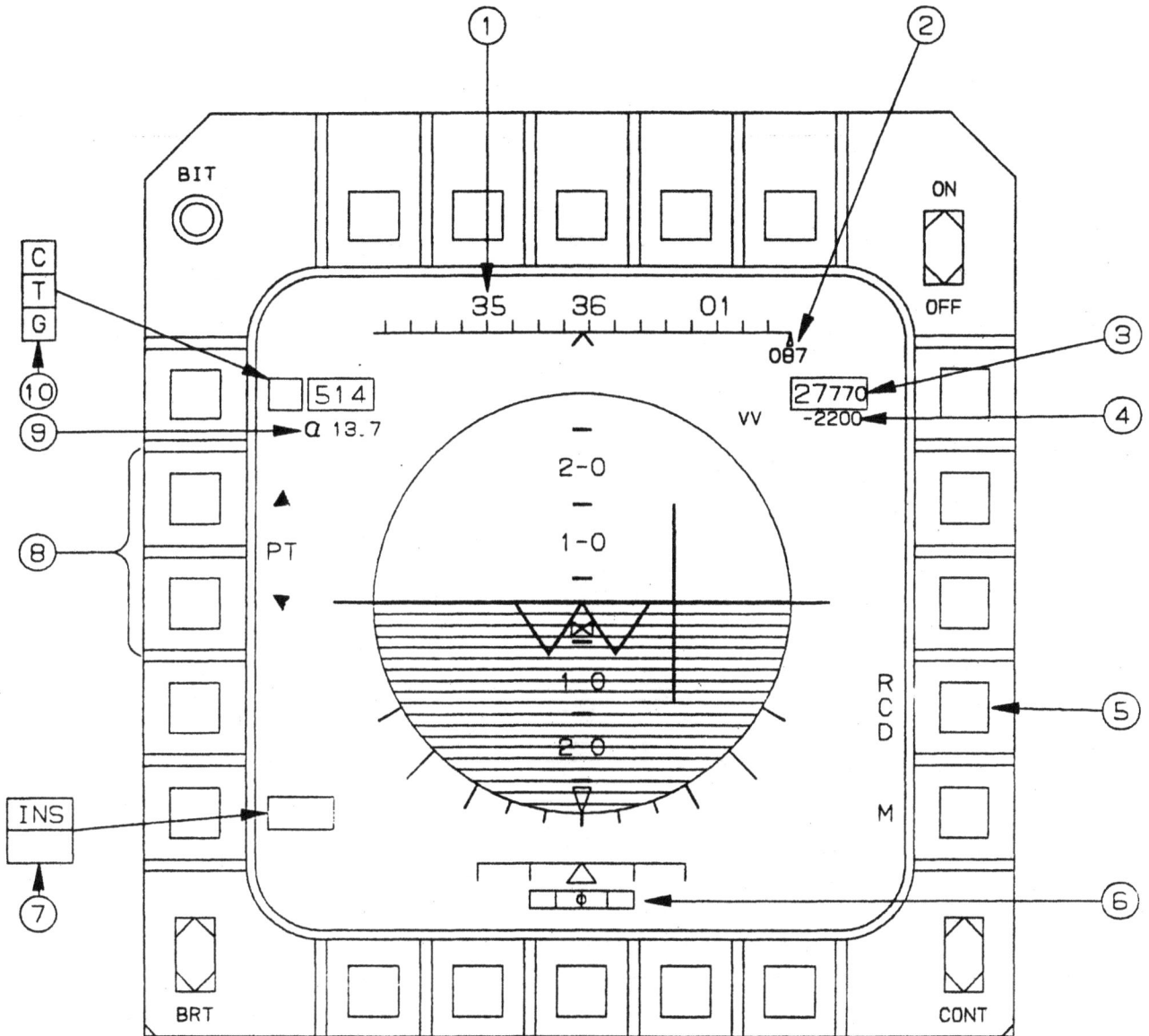

1. HEADING SCALE
2. COMMAND HEADING (OFF SCALE)
3. ALTITUDE
4. VERTICAL VELOCITY
5. RECORD BUTTON
6. TURN AND SLIP INDICATORS
7. ATTITUDE SOURCE INDICATION
8. PITCH REFERENCE BUTTONS
9. ANGLE OF ATTACK INDICATOR
10. SPEED INDICATOR

A23-017

**Figure 1-43**

## ELECTRONIC ADI SYMBOLS

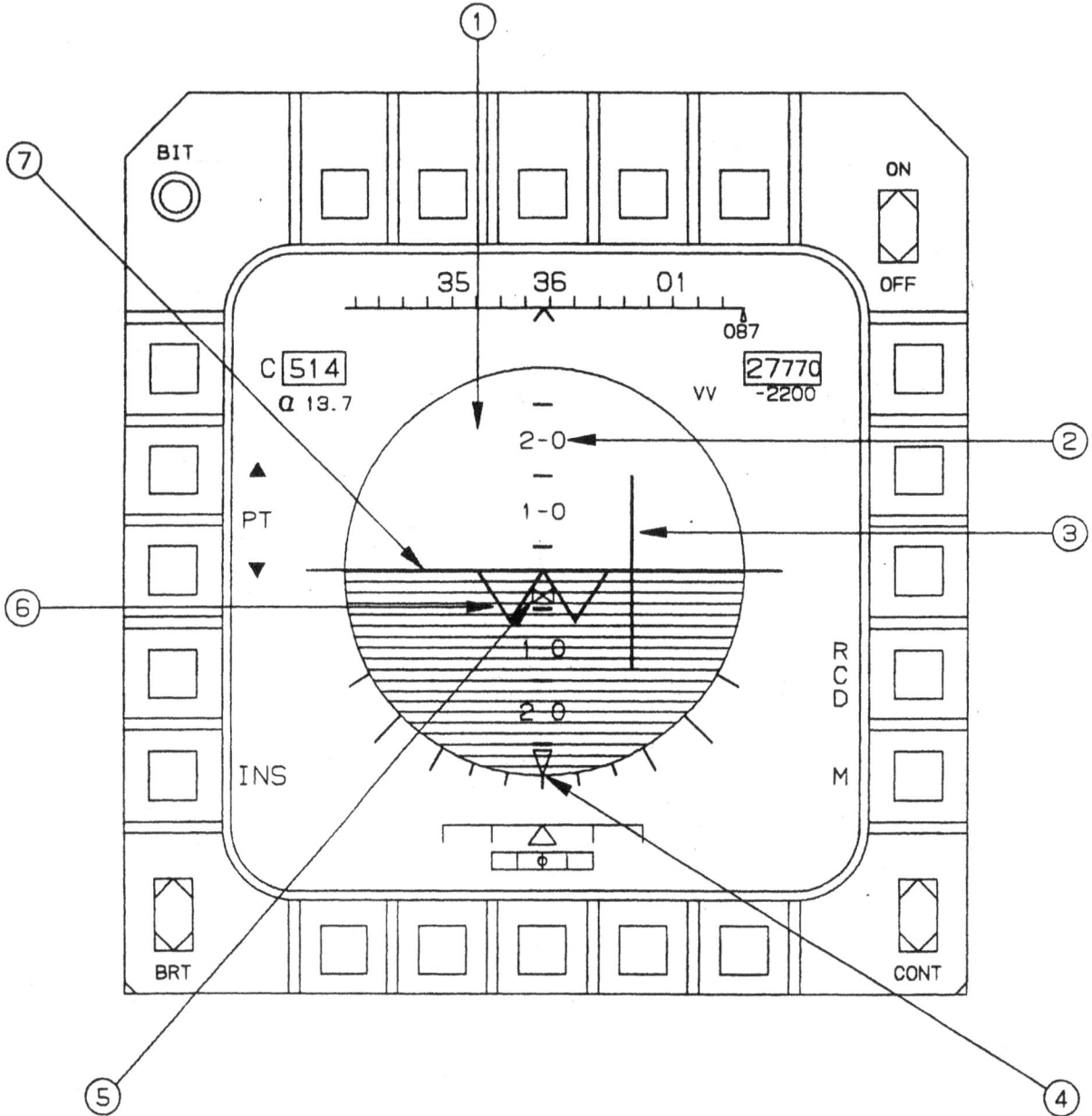

1. ATTITUDE REFERENCE WINDOW
2. PITCH SCALE
3. BANK STEERING BAR
4. BANK SCALE AND POINTER
5. FLIGHT PATH MARKER
6. AIRCRAFT
7. HORIZON

A23-018

Figure 1-44

## Attitude Reference

Attitude reference is similar to a conventional attitude indicator, using inertial navigation system (INS) reference data. Attitude source data in the lower left corner of the display indicates normal (INS) or invalid (INS OFF) INS status.

The attitude reference window is the center of the display. The area above the horizon is blue, below the horizon is yellow.

Pitch scale bars are at five degree increments above and below the horizon, with numbers every ten degrees. The flight path marker moves up and down the pitch scale showing vertical flight path in relation to the horizon.

Bank angle tic marks are at 0, 10, 20, 30, 45, 60 and 90 degrees.

The aircraft symbol is fixed in relation to the MPCD. Display adjust buttons move the horizon line up or down.

### Electronic ADI Heading

A white heading scale and white pointer provide heading data. The heading scale scrolls past the pointer, with tic marks every two degrees. The scale shows 15 degrees each side of the heading. Heading is from the INS.

The command heading marker is set from the electronic HSI. If the command heading is not within the range shown on the heading scale, the marker parks at one end of the scale, indicating the direction to turn, and a digital readout of the command heading appears under the marker.

The turn and slip indicator is below the attitude reference window. Tic marks on the turn rate scale are for 1-minute and 2-minute turns.

### ADI Flight Data

Altitude is based on the altimeter setting entered in the UFC. Altitude readout is rounded to 10 foot increments. The vertical velocity includes a minus sign for negative values and is rounded to hundreds of feet per minute.

Speed information is identified as calibrated airspeed (C), true airspeed (T), or groundspeed (G). Selection is through the UFC.

The Greek letter alpha identifies angle-of-attack data.

The bank steering bar reflects navigation guidance selected on the electronic HSI.

## Invalid ADI Data

If data is invalid, a red word OFF replaces the data. A red X superimposed on the attitude reference window indicates invalid INS data. Heading, vertical velocity, and turn rate data are invalid.

## ELECTRONIC HSI

The electronic HSI presents navigation data (figure 1-45). The display graphically presents heading, track, course, destination, and TACAN information. Related navigation data is in text. Steering information is available for INS (NAV), ground track (GT), or TACAN (TCN) guidance by pressing the appropriate button.

Valid TACAN and INS information is on the HSI with any guidance selection. Symbols and readouts show relative positions of a TACAN station and a selected steer point, along with bearing, distance, and ETA. Course deviation indication is available when TACAN is selected. TACAN data is green. INS data is blue.

### HSI Heading

A compass rose rotates around the fixed aircraft symbol. Heading numbers and symbols at 30-degree intervals remain upright as the compass rose rotates during turns. The orange heading marker (captain bars) rotates with the compass rose.

The heading marker shows the command heading (CMD HDG) to make good the course selected for ground track or INS guidance. With TACAN guidance selected, reference heading (HDG) is entered with heading set buttons, independent of course selection. Reference heading or ground track can also be entered through the UFC scratchpad.

Arrowheads appear in adjacent fields when heading or course select buttons are active. An arrow pointing up increases the related value. With ground track guidance selected, course selection is available. TACAN guidance selection permits heading and course entries. With INS guidance, course information depends on steer point selection from the UFC.

### HSI Display Range

The range shown is from the aircraft symbol to the inner edge of the compass rose. The buttons adjacent to the arrows increase or decrease the depicted range. Selectable ranges are 10, 20, 40, 80, and 160 nautical miles.

## *ELECTRONIC HSI*

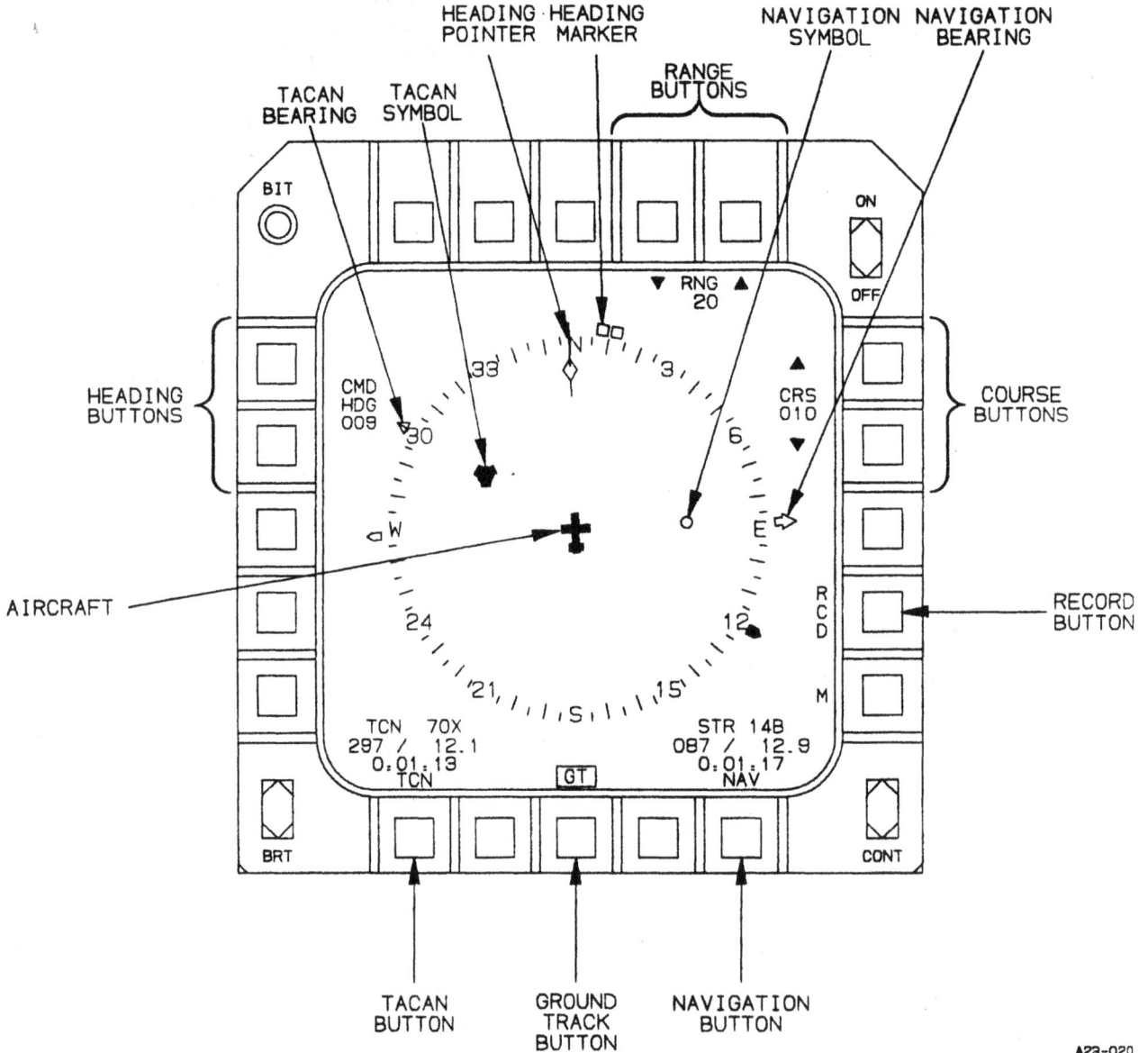

Figure 1-45

**NORTHROP/MCAIR YF-23A**
**COMPETITION SENSITIVE**
**UNCLASSIFIED**

A23-020

## AVIONICS INTERFACE

Data communications among avionics systems is through multiplex (MUX) buses and discrete connections (figure 1-46). The central computer is the avionics bus controller. Avionics interface units and the multipurpose display processor provide additional avionics control and computation functions. Input to avionics is through throttle and control stick switches, MPCD, and UFC.

System and component redundancy provides backup for essential avionics functions. Automatic redundancy management affects all operations of the integrated avionics system.

### Central Computer

The central computer (CC) monitors and controls MUX bus components and functions, and performs computations for controls and displays, weapons, and navigation systems. The computer monitors the condition of components and controls the use of systems redundancy. Using data from VMS, INS, and radio navigation system, the computer provides guidance information for flight instrument displays.

The multipurpose display processor backs up the central computer in case of computer or dual generator failure. Pressing the computer reset button, labeled CC on the INS control panel, reinitializes the computer after malfunction or failure (figure 1-47).

The ground power panel central computer power switch controls power from left main ac bus 1 for the central computer (figure 1-48).

### Multipurpose Display Processor

The multipurpose display processor (MPDP) controls and supports MPCD and the HUD and is the backup avionics bus controller. The display processor automatically takes over essential central computer functions. Display processor redundancy includes a secondary driver for the HUD.

The ground power panel multipurpose color display processor power switch controls power from right main ac bus 1 for the display processor.

### Avionics Interface Units

Two avionics interface units (AIU) control data and electrical power for controls and displays and for communications and navigation radios. The interface units also pass signals from throttle and stick switches, control initiated BIT, report avionics status, process master caution light and tone signals, and prepare cautions and advisories for the MPCD. The

## AVIONICS INTERFACES

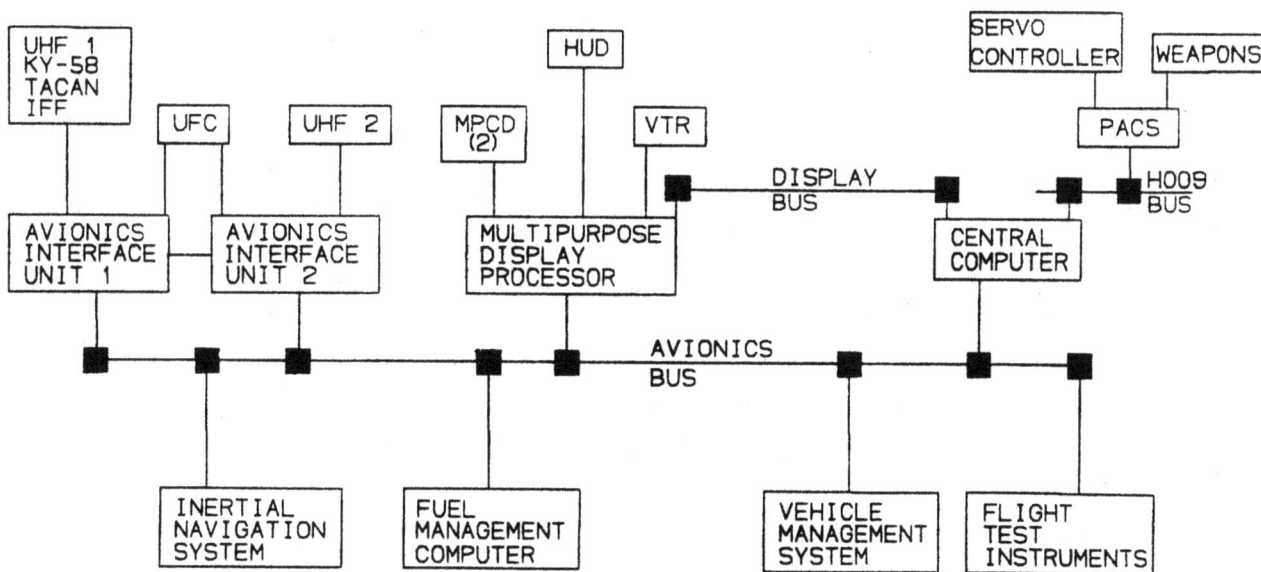

HUD - HEAD-UP DISPLAY
MPCD - MULTIPURPOSE COLOR DISPLAYS
UFC - UPFRONT CONTROL
VTR - VIDEO TAPE RECORDER

A28-022

**Figure 1-46**

## CENTRAL COMPUTER RESET BUTTON

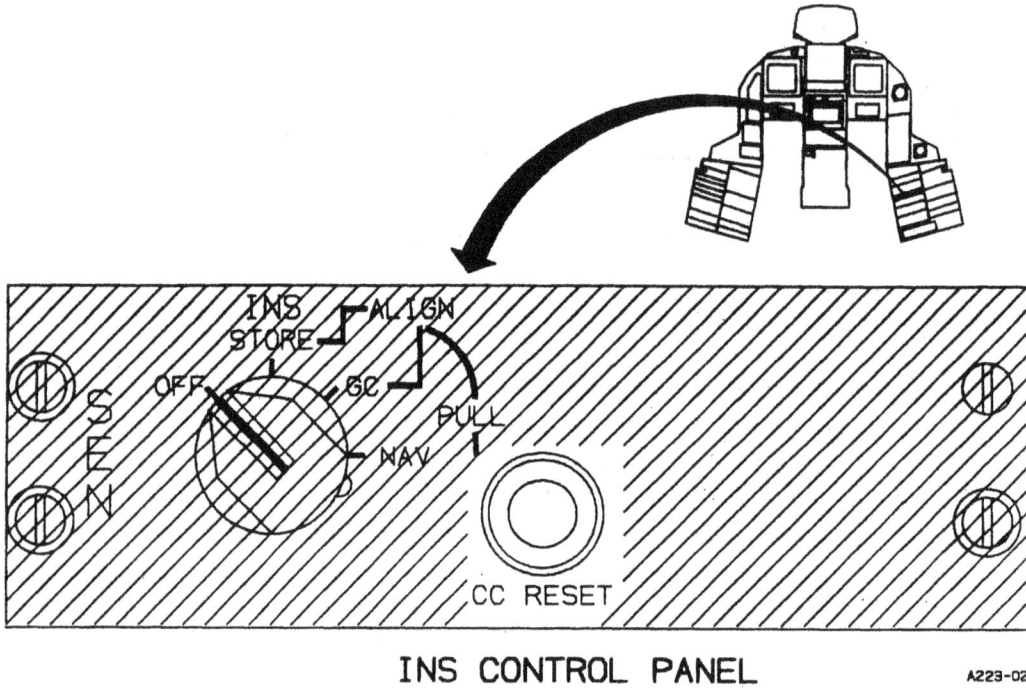

INS CONTROL PANEL

A223-023

*Figure 1-47*

## GROUND POWER PANEL

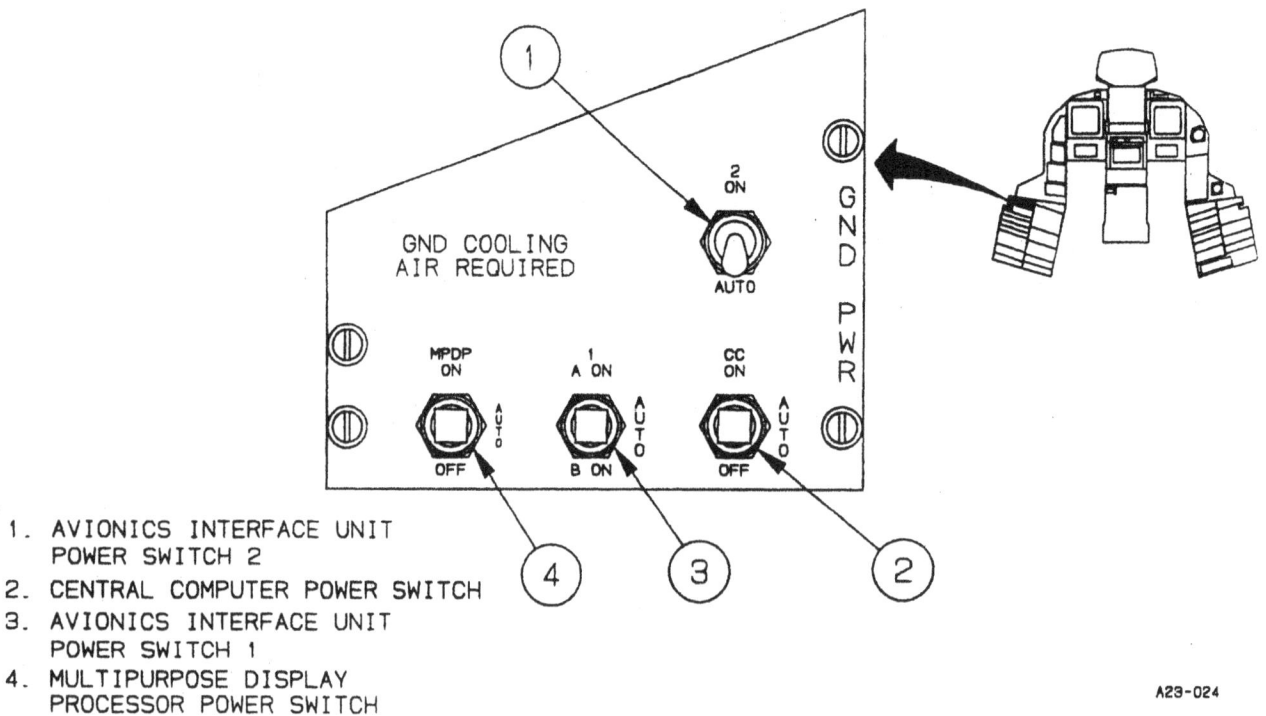

GND COOLING
AIR REQUIRED

1. AVIONICS INTERFACE UNIT
   POWER SWITCH 2
2. CENTRAL COMPUTER POWER SWITCH
3. AVIONICS INTERFACE UNIT
   POWER SWITCH 1
4. MULTIPURPOSE DISPLAY
   PROCESSOR POWER SWITCH

A23-024

*Figure 1-48*

interface units provide redundancy support for each other.

Interface unit 1 has two processors (1A and 1B) and unit 2 has one processor. Backup is available for all processor functions except:

| | |
|---|---|
| Processor 1A | Caution and advisory processing: |
| | • Normal control of UHF 1 and IFF, |
| | • KY-58 and TACAN operation, |
| | • Avionics BIT. |
| | |
| Processor 2 | UHF 2 control. |

Emergency operation of UHF 1 and IFF is available if processor 1A fails. If the IFF master switch on the intercom panel is in the emergency position, IFF emergency code transmission (all applicable modes) and UHF 1 Guard receiver are available.

The multipurpose display processor power switch controls power from the left essential ac bus for avionics interface unit 1. Switch 1, position B, on the ground power panel controls interface unit 2 power from right main ac bus 1 and the left main dc bus.

## BACKUP AVIONICS OPERATION

Power source and functional redundancy automatically provide essential functions. Either generator supports all avionics except IFF mode 4 and the programmable armament control set. Failure of major components is indicated by caution, advisory, or BIT messages, or by changes in controls and displays configuration.

### Multipurpose Display Processor Bus Control

If the multipurpose display processor assumes bus control, the flashing signal STANDBY replaces the display on the left MPCD and a modified ADI is on the right MPCD. Other displays are not available. Missing ADI features are: angle-of-attack and command heading indications, and display adjust, attitude source, and record buttons.

The display processor continues latitude and longitude computations. The modified ADI has buttons for selecting TACAN or navigation guidance, with TACAN and steer point information in the lower left and right corners. Data for two steer points can be entered through the UFC. Course select buttons are active if TACAN guidance is selected.

Cautions and advisories appear on the modified ADI. Any movement of the display selector alternately removes and restores the messages. The display processor does not back up central computer fuel computations. The MPCD caution FUEL FAIL and advisory BINGO are not supported. Caution messages FUEL ASYM and TANK PRESS appear continuously.

See UFC Controls and Menus, this section, for UFC availability with the display processor as MUX bus controller.

### INS Failure

The INS provides position, attitude, slip, and velocity data to controls and displays. The VMS provides airspeed and backup slip information. If INS data is not available, a red X overlays the ADI attitude reference window and affected information is missing or shown as off on ADI and HSI. Ground speed and vertical velocity are not available. TACAN information is available.

## VIDEO TAPE RECORDING

Video tape recording for flight test is available using the record (RCD) button on electronic ADI and HSI, tactical situation, fuel, or engine displays. See Video Tape Recording, this section.

Video tape recording of the HUD is normally continuous when the flight test function control panel is set for avionics data recording. Time-of-day readout at lower left of the HUD field of view indicates HUD recording. Video tape recording of MPCD is selectable from the electronic ADI and HSI, and from tactical situation, engine, or fuel displays, using the record (RCD) button. When active, the record button starts or stops recording of that MPCD. A time readout replaces the record label during MPCD recording. When an MPCD is selected for recording, any MPCD display can be selected and recorded.

Split-screen recording replaces normal HUD recording during MPCD recording. If both MPCD are selected for recording, temporary HUD recording can be selected with control stick trigger or weapon release switch. The HUD replaces the left MPCD.

Squeezing the trigger to the first detent starts HUD recording and places visual and audible event markers. Recording continues while the trigger is in either detent, and for ten seconds after release from the second detent. Pressing the weapon release switch starts recording and places event markers. HUD recording continues for ten seconds after release of the switch.

Tape recording includes voice communications and audible warning, caution, and advisory signals.

## BUILT-IN TEST

Built-in test (BIT) monitors all systems except exterior lights. The central computer coordinates BIT, with avionics interface unit 1A controlling test of avionics. Initiated test is available through control panels and the MPCD BIT displays. BIT displays permit initiated BIT of MUX bus components. See system descriptions for other BIT applications.

### Built-in Test Displays

The BIT displays permit initiated BIT of components, individually or in selected groups, and show the results of initiated or continuous test. BIT displays also provide access for detailed maintenance test and for identification of failures at the circuit card level.

The top level BIT display, selected from the MPCD menu, provides options for ground and airborne initiated BIT, and shows initiated BIT status and failures detected by BIT (figure 1-49). Units currently in initiated BIT are identified as IN TEST in the upper portion of the display area. Failures triggering the advisory AV BIT are identified in the lower portion. An asterisk next to a component identifier indicates further information is available on an associated detail BIT display.

BIT DISPLAY BUTTONS

The buttons labeled MAINT (maintenance) and DETAIL select subordinate BIT displays. Other buttons initiate BIT. Labels identify the available BIT options on the ground or airborne. Labels are AUTO (automatic BIT), VMS, INS, AIU (avionics interface units), IFF, TCN/COMM (TACAN and communications radios), FUEL LOW, FMC (fuel management computer), PACS (programmable armament control set), AIM (AIM-120), DSPL (displays), and UFC.

The label AIM normally appears. The label PACS appears for ground use, with ground BIT consent switch and no weapons on board.

INS BIT is for maintenance use.

During initiated BIT the selected label remains, other option labels disappear, and buttons with labels STOP and ESCAPE are active. The stop button terminates initiated BIT. If a test process does not run properly, pressing the escape button resets BIT functions in the central computer, permitting test of other systems.

### Automatic BIT

Automatic BIT is a series of tests, lasting approximately two minutes, available in flight or on the ground. On the ground, automatic BIT includes avionics interface units, IFF, TACAN, communications, fuel low, fuel management computer, armament (PACS or AIM), displays, and UFC tests. All except displays are tested in flight. The aircraft must be motionless for ground test.

### VMS Initiated BIT

VMS initiated BIT is a preflight function, selected by simultaneously pressing the BIT consent switch on the VMS control panel and the VMS button on the BIT display. The VMS computer must be operating and the aircraft must be stationary, with landing gear and gear handle down and throttles in idle. An error message identifies an improperly configured system. If BIT is interrupted, INCOMPLETE appears. The message INCOMP-HYD indicates hydraulic pressure was not applied to actuators during test.

### TACAN and Communications BIT

The button labeled TCN/COMM selects test of TACAN, UHF radios, and intercommunication system. Tests of TACAN and radios are concurrent. Test of the intercom system is not interrupted by the stop button.

### Fuel Low BIT

Fuel sensors are tested by fuel low BIT. Successful test clears the messages L FEED NO-GO and RIGHT FEED NO-GO, which appear when power is applied to the fuel management computer.

### Display and UFC BIT

During display test, MPCD and HUD initially are blank, then test patterns appear. Pressing the stop button on the right MPCD replaces test patterns with BIT displays and initiates UFC BIT. Test of the UFC is menu-driven.

### Maintenance BIT Display

The BIT display button labeled MAINT selects the maintenance BIT display (figure 1-50), providing access to functions, some available in flight but primarily for ground use. Failure messages do not appear on displays selected from the maintenance BIT display. Buttons labeled VMS, AUDIT, and INS are for ground tests and access to computer memory.

## BUILT-IN TEST DISPLAY

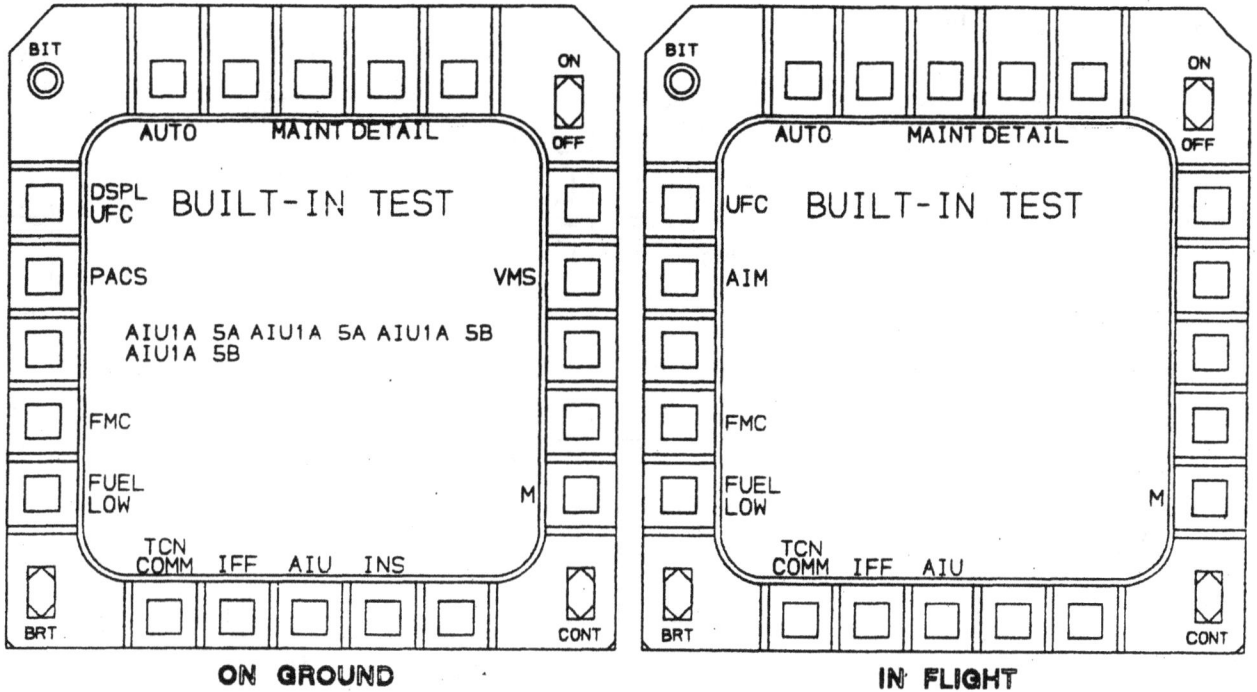

ON GROUND

IN FLIGHT

TOP LEVEL BIT DISPLAYS

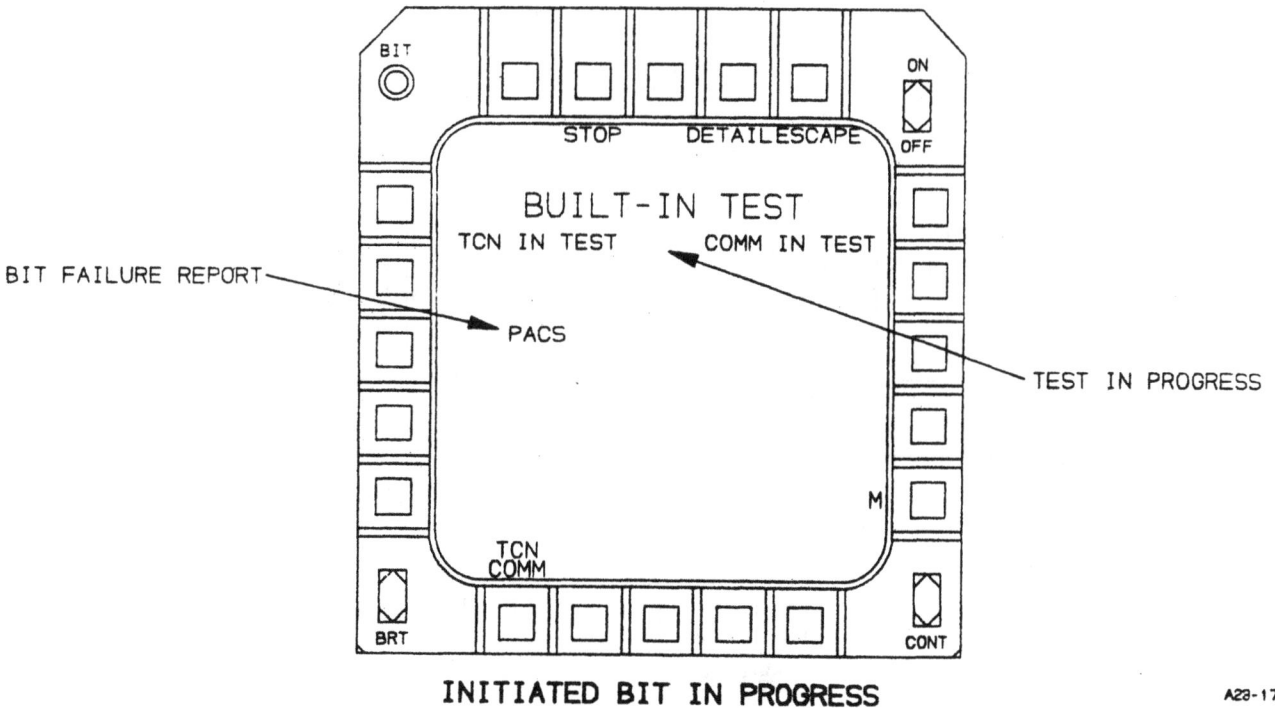

INITIATED BIT IN PROGRESS

A23-176

Figure 1-49

## MAINTENANCE BIT DISPLAY

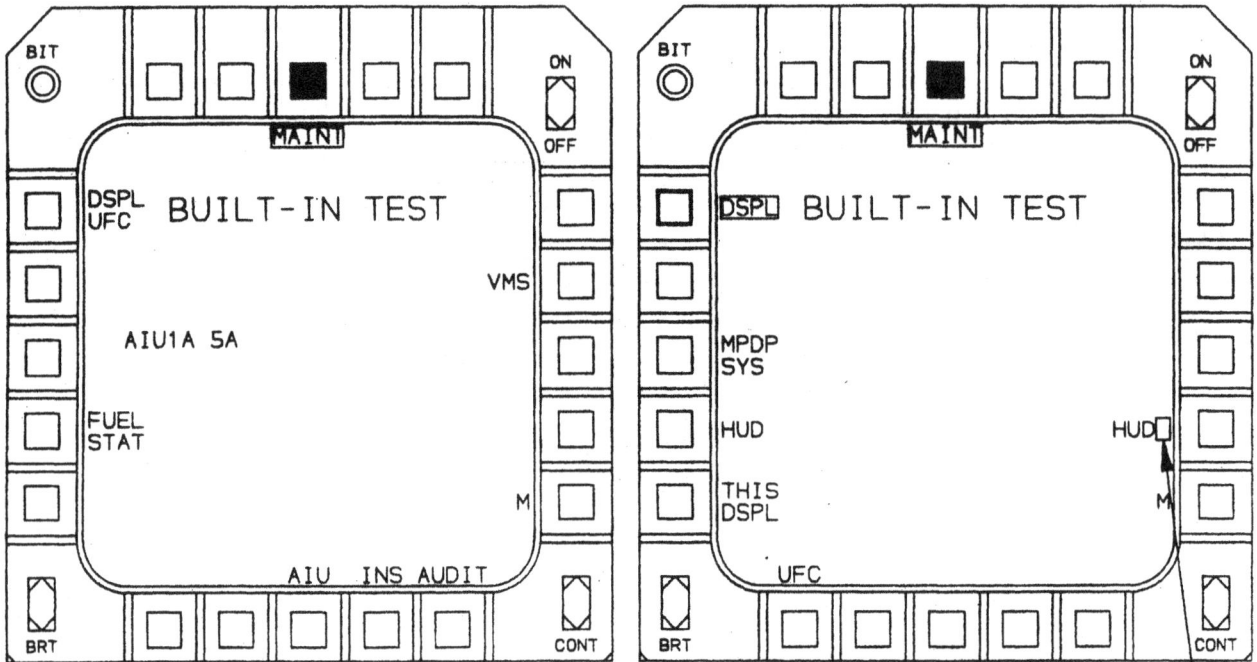

MAINTENANCE BIT DISPLAY

DISPLAYS MAINTENANCE
BIT DISPLAY

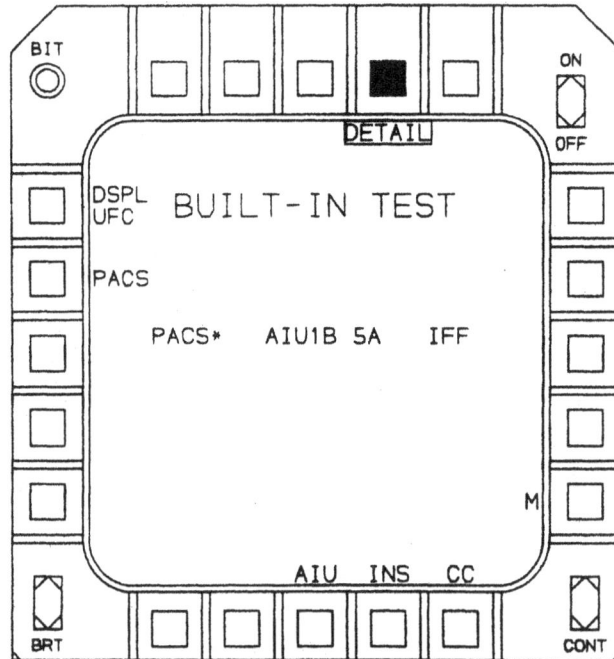

DETAIL BIT DISPLAY

A28-177

*Figure 1-50*

DSPL
UFC

Selects displays maintenance BIT display, giving access to test of controls and displays. On the displays maintenance BIT display, the button labeled MPDP SYS selects ground test of multipurpose display processor and color displays. Inflight test is available for all except the display processor. Tests of MPCD and HUD must be terminated with the stop button.

If a HUD driver fails in the display processor, the HUD driver button alternately selects the primary driver (HUD label with boxed P) or the secondary (boxed S).

AIU

Selects avionics interface unit override display. If there are faults or failures in both interface units, the central computer assigns redundant functions to one interface unit. The override display label AIU1 or AIU2 indicates the active interface unit. The adjacent button selects the other unit.

FUEL
STAT

Selects fuel status BIT display, for monitoring and controlling fuel system components on the ground and for monitoring the system in flight.

**Detail BIT**

The detail BIT display, selectable from the top level BIT display when a failure indication includes an asterisk, identifies failures to the circuit card level. Detail BIT information is available for controls and displays, armament, INS, avionics interface units, and the central computer. The display is available in flight and on the ground.

# COMMUNICATION SYSTEMS

The communications system includes the intercommunication system (ICS), two UHF radios, support antennas (figure 1-51), KY-58 secure voice, and the IFF.

# INTERCOMMUNICATION SYSTEM

The ICS interfaces with navigation radios, communications, warning, caution, advisory, and other systems through the avionics interface unit. The ICS consists of the integrated communications panel and the inboard throttle microphone switch.

The intercommunication panel controls voice communication and receiver volumes except the UHF radios. (figure 1-52).

## UHF Antenna Switch

The UHF antenna switch provides two UHF radio options.

| | |
|---|---|
| 1 UPR/2 LWR | Upper antenna selected for UHF 1. Lower antenna selected for UHF 2. |
| 2 UPR/1 LWR | Upper antenna selected for UHF 2. Lower antenna selected for UHF 1. |

## Intercommunication Volume and Weapons Volume

The inner knob controls the intercommunication system volume. The outer knob controls the weapons aural tone volume.

## TACAN Volume

The outer knob controls the TACAN identifier volume. The inner knob is not used.

## INTERCOM Switch

The intercom switch controls microphone transmissions.

| | |
|---|---|
| RAD ORIDE | Attenuates radio reception volume and improves intercom communications. The unsafe landing gear warning **Tone** and voice warning messages operate normally. |
| ON | Provides HOT MIC communications directly with ground crew. |

| | |
|---|---|
| OFF | Turns intercom off. Radio transmission/reception is unaffected. |

## Warning Tone Silence Button

Silences warning tones. See Warnings, Cautions, and Advisories, this section.

## IFF Master Switch

Selects the desired output for the IFF transponder.

| | |
|---|---|
| LOW | The IFF operates at reduced sensitivity. Mode 4 is not affected. |
| NORM | Selects normal IFF |
| EMERG | The IFF transmits Mode 3, 7700. |

## Mode 4 Reply Light

Lights white when Mode 4 responds to an interrogation.

## Mode 4 Switch

Provides Mode 4 reply notification.

| | |
|---|---|
| LIGHT | Enables reply light to respond during interrogation. |
| AUDIO REC | Enables both the Mode 4 reply light and an audio tone to respond during interrogation. |
| OFF | Both the Mode 4 audio tone and reply light functions are off. |

## Mode 4 Code Switch

Provides Mode 4 options.

| | |
|---|---|
| B | Selects Mode 4/B reply. |
| A | Selects Mode 4/A reply. |
| OUT | Mode 4 reply function is off. |

## Mode 4 Crypto Switch

Controls Mode 4 hold or zero function, and a KY-58 code zero function.

| | |
|---|---|
| HOLD | Mode 4 code held when IFF turned off. Switch is spring-loaded out of HOLD position. |

## ANTENNA LOCATIONS

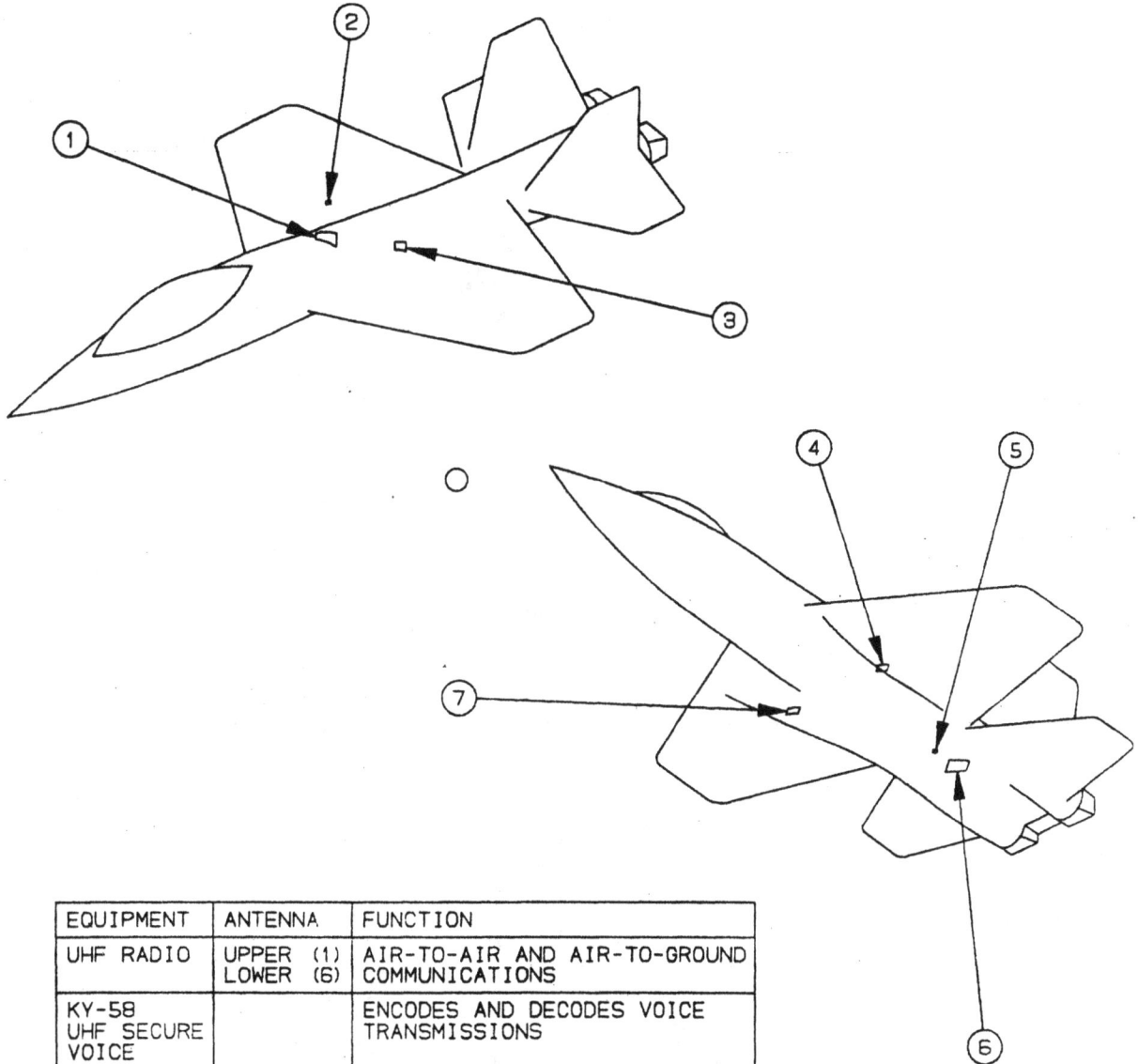

| EQUIPMENT | ANTENNA | FUNCTION |
|---|---|---|
| UHF RADIO | UPPER (1)<br>LOWER (6) | AIR-TO-AIR AND AIR-TO-GROUND COMMUNICATIONS |
| KY-58<br>UHF SECURE<br>VOICE | | ENCODES AND DECODES VOICE TRANSMISSIONS |
| IFF | UPPER (1)<br>LOWER (4) | AIRCRAFT IDENTIFICATION TO INTERROGATING RADAR |
| C-BAND<br>BEACON | UPPER (2)<br>LOWER (5) | FLIGHT TEST RANGE |
| L-BAND<br>BEACON | UPPER (3)<br>LOWER (7) | FLIGHT TEST TELEMETRY |
| TACAN | (6) | ●BEARING AND DISTANCE FROM GROUND TACAN AND SUITABLY EQUIPPED AIRCRAFT<br>●DISTANCE INFORMATION FROM SIMILIARLY EQUIPPED AIRCRAFT |
| INTERCOMMUNICATION SYSTEM | | GROUND CREW COMMUNICATION AND VOLUME CONTROL |

A29-152

*Figure 1-51*

# INTERCOMMUNICATION PANEL

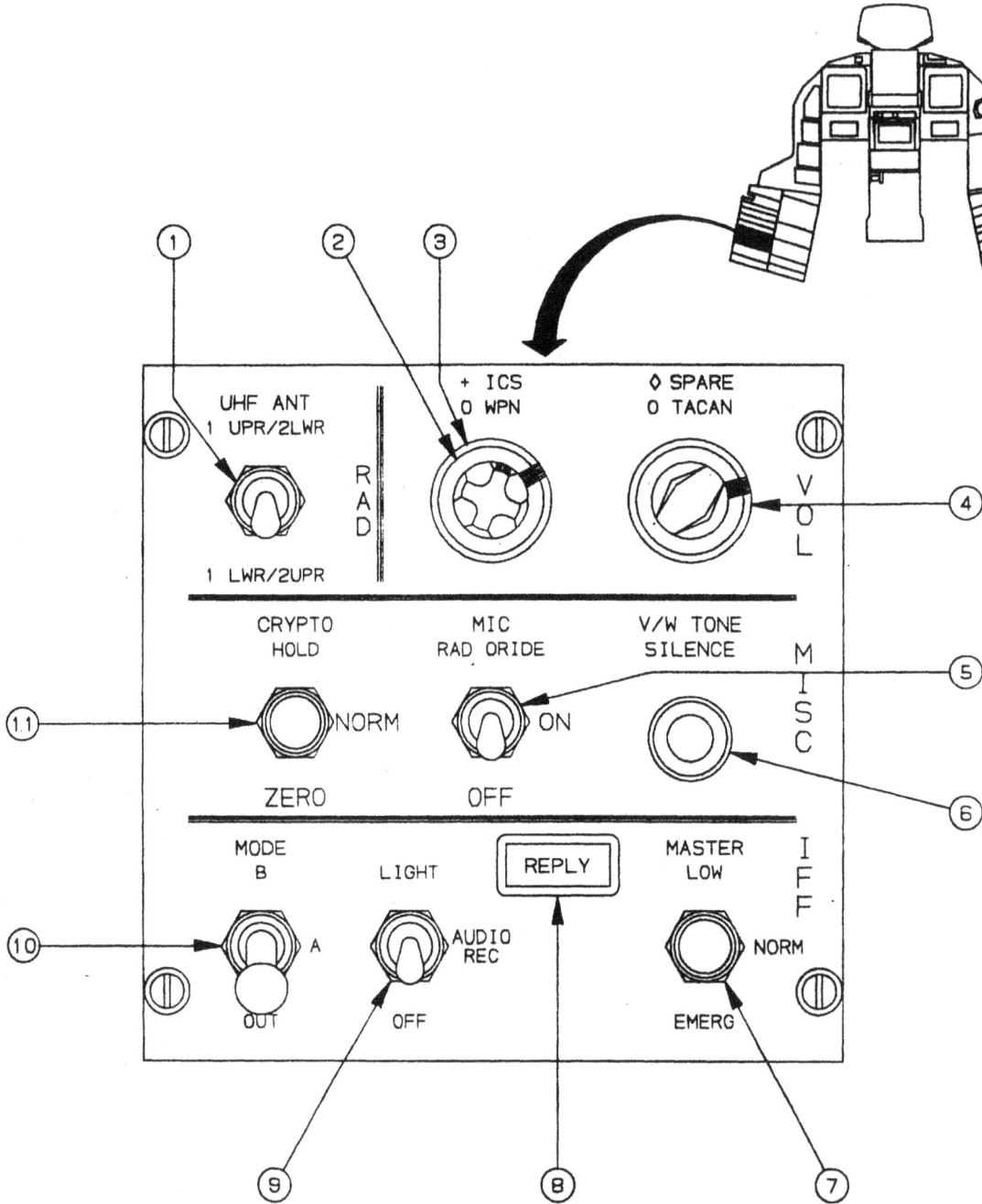

1. UHF ANTENNA SELECTOR SWITCH
2. INTER-COMMUNICATIONS VOLUME
3. WEAPONS VOLUME
4. TACAN VOLUME
5. INTERCOM SWITCH
6. WARNING TONE SILENCE BUTTON
7. IFF MASTER SWITCH
8. MODE 4 REPLY LIGHT
9. MODE 4 REPLY SWITCH
10. MODE 4 CODE SWITCH
11. MODE 4 CRYPTO SWITCH

A23-146

*Figure 1-52*

| | | | |
|---|---|---|---|
| NORM | Enables Mode 4 operation when aircraft is powered. | R1<br>UHF frequency | Changes frequency of preset channel from scratchpad entry. |
| ZERO | Zeroizes Mode 4 and KY-58 codes. Switch is spring-loaded out of ZERO position. | L2<br>SQUELCH | Cycles squelch on (*) and off. |

### Microphone Switch

The inboard throttle microphone switch is spring-loaded to the center receive position. Pushing the switch forward transmits on UHF 1. Pushing the switch aft transmits on UHF 2.

# UHF RADIOS

The UHF radios transmit/receive both encrypted and plain voice radio communications and monitor UHF guard frequency (243.0 MHz). Each radio operates on UHF frequencies 225.000 to 399.975 MHz, in 0.025 MHz increments. Twenty preset frequencies may be stored by the upfront control (UFC), or the UFC scratchpad can be used to enter a manual frequency. The UHF radios also have antijam (have quick) capability. When power is removed, each radio stores the last selected frequency.

Channels and frequencies for the UHF radios appear on all UFC menus and displays. UHF 1 readouts are on the left side of each menu, and UHF 2 readouts are on the right side. The volume control for UHF 1 is on the left side of the UFC and labeled R1. UHF 2 volume control is on the right side and is labeled R2.

### GUARD SELECTION

UFC guard receive is selected on/off by pressing the shift (SHF) button and the appropriate UHF tuning (GREC) key. The left UHF tuning key controls UHF 1 guard and the right UHF tuning key controls UHF 2 guard. A **G** next to the manual frequency readout indicates the guard receiver is active.

Transmit on guard radio frequency by rotating the UHF preset channel selector until a **G** appears next to the preset channel readout and pressing the microphone switch.

### UHF MENUS

The UFC UHF 1 or UHF 2 menu provides control of UHF power, antijam functions, and squelch (figure 1-53).

| | | |
|---|---|---|
| L1<br>U1 or<br>U2 | | Cycles radio on (:) and off. |

| | |
|---|---|
| R2<br>NB | Narrow band function for antijam. Steps to wide band. |
| WB | Wide band function for plain or encrypted transmissions. Steps to narrow band. |
| L3<br>AJ | Cycles antijam on (*) and off. Selects narrow band. |
| R3<br>TOD | Enables (*) radio to receive time-of-day (TOD) signal for antijam operation. |
| L5 | Tunes manual frequency on scratchpad into UHF 1. Press again to return to original frequency. |
| R5 | Tunes manual frequency on scratchpad into UHF 2. Press again to return to original frequency. |

### UHF FREQUENCIES AND PRESET CHANNELS

An asterisk (*) appears next to the frequency readout indicating manual frequency selection for the appropriate radio. If the asterisk appears next to the preset channel readout, then the radio operates on the preset channel. To change from the preset channel to a manual frequency, press the UHF tuning (C M) key. The asterisk moves between the two readouts.

To change a stored preset channel frequency, select the menu and the preset channel of the UHF radio to be changed. Enter a valid UHF radio frequency in the scratchpad and press R1 adjacent to the old frequency.

A **C** next to the preset channel readout indicates the KY-58 secure voice system (cipher) is active. An **A** next to the preset channel readout indicates the antijam function is active.

## *UHF 1 AND 2 MENUS*

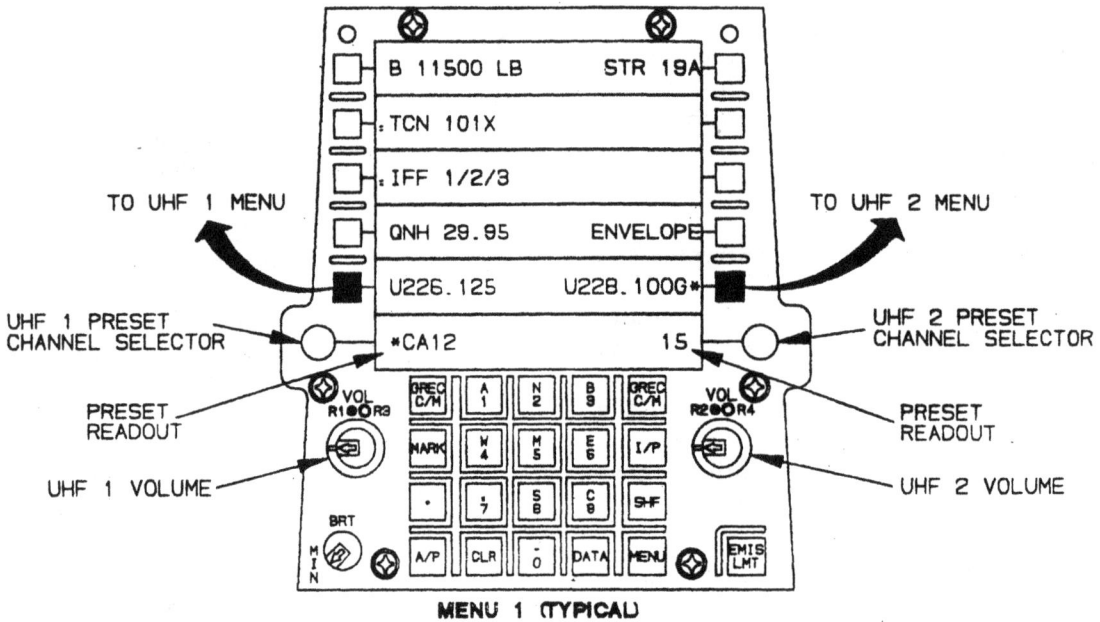

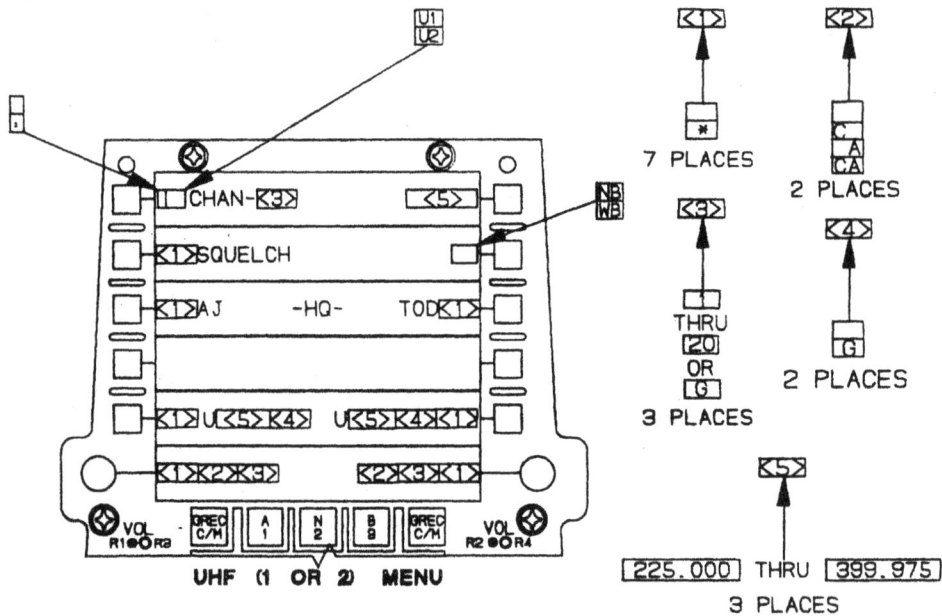

Figure 1-53

# UHF SECURE VOICE

The KY-58 secure voice system (figure 1-54) encodes and decodes either UHF radio. Only one radio can transmit on secure voice at a time.

## MODE SELECTOR

The mode selector provides selection of plain or encrypted UHF transmissions and the loading of encryption variables.

| | |
|---|---|
| P | Transmissions are plain (not encrypted). Plain transmissions are received. |
| C | UHF transmissions for encryption are controlled through the UFC. Plain or encrypted transmissions are received. |
| LD | Not used. |
| RV | Not used. |

## FILL SELECTOR

The fill selector selects memory locations for coding transmissions/receptions and for erasing loaded codes. Selector must be pulled out before rotating.

| | |
|---|---|
| 1-6 | Access to indicated storage registers. |
| Z 1-5 | Zeroizes registers one through five. |
| Z ALL | Zeroizes all six registers. |

## FILL CONNECTOR

Connection for attaching the encoding equipment.

## POWER SELECTOR

The power selector provides power to the KY-58.

| | |
|---|---|
| OFF | Unit is in standby. |
| ON | Enables unit operation. |
| TD | Not used. |

## CODE DATA CARD

Coded registers information is noted on data card.

## *UHF SECURE VOICE*

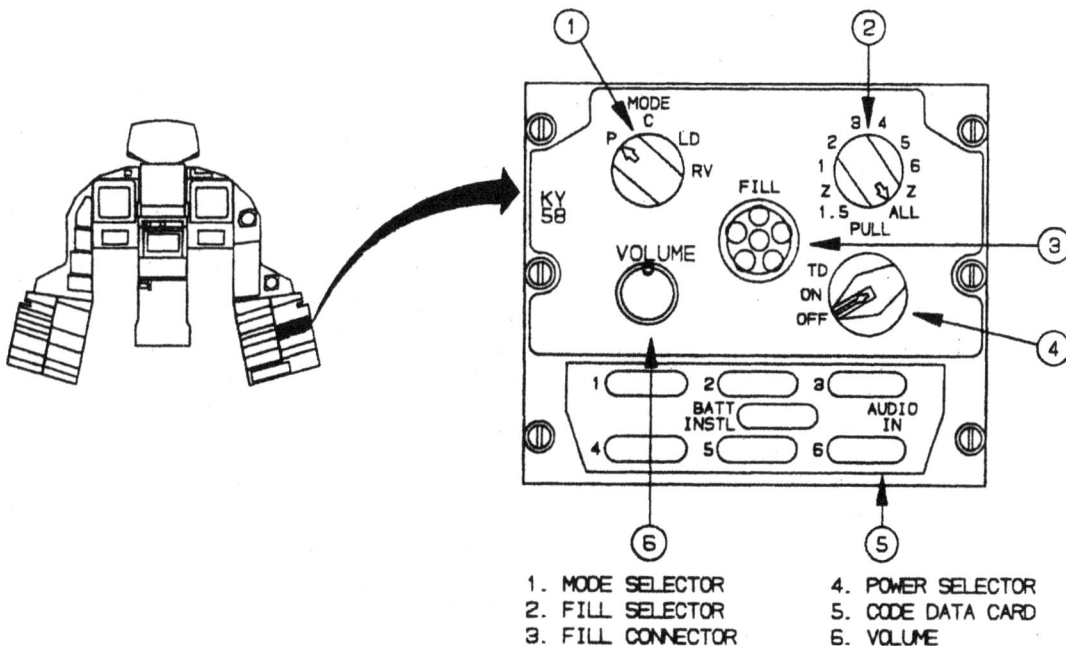

1. MODE SELECTOR
2. FILL SELECTOR
3. FILL CONNECTOR
4. POWER SELECTOR
5. CODE DATA CARD
6. VOLUME

A23-153

**Figure 1-54**

## VOLUME

Adjusts the audio level of the KY-58. Volume should be full clockwise, permitting adjustment by use of the UFC volume controls.

# UP FRONT CONTROL OF KY-58

When operating the secure voice from the up front control, the KY-58 panel switches should be preset:

- Mode Selector Knob -- Not functional
- Fill Selector -- Any setting except Z (zeroize).
- Power Selector -- Set to ON.
- Volume -- Set to maximum.

### KY-58 MENU

The UFC KY-58 menu provides control of the UHF radio secure voice system. Select the KY-58 menu by R1 on menu 2 (figure 1-55).

| L1 OPR | Applies (:) KY-58 operation power. Receive variable disabled. |
|---|---|
| R1 RV | Not used, should be off. |
| L2 CIPHER | Transmissions are encrypted. Steps to plain. |
| PLAIN | Transmissions are not encrypted. Steps to cipher. |
| R2 DELAY | Not used, should be off. |
| L3 BB | Secure voice in baseband is not used. Step to diphase. |
| DP | Secure voice is in diphase. Use DP. |
| R3 FILL | Secure voice is available if variables are loaded. If the scratchpad contains a number 1 through 6, the corresponding register is entered for transmission. Steps to KEY FILL if pressed with an empty scratchpad. |
| KEY FILL | Permits variable loading. Steps to fill if pressed with an empty scratchpad. If the scratchpad contains a number 1 through 6, the corresponding |

register is entered for loading. To fill, key mic.

| L4 U1 | Selects (*) secure voice for UHF 1. |
|---|---|
| R4 U2 | Selects (*) secure voice for UHF 2. |

### HAVE QUICK

The have quick system provides frequency-hopping antijam for UHF radio by changing channels or frequencies at variable times. All radios in a particular operation must have the same word-of-day (WOD), time-of-day (TOD), net number, and frequencies stored in allocated preset channels. For training nets, channel 20 is used to store the WOD and channels 14 thru 20 are used to store the preset frequencies.

Each radio contains a clock, memory circuits, and a real time code generator. This enables the radios to change frequencies at variable times. The WOD provides the hopping pattern and the TOD signal synchronizes the internal clock to other radios in the net.

Each radio accepts the first TOD received on any channel or frequency in use after power is applied. The TOD signal is heard in the headset as a short burst of varying tones followed by a steady tone. The steady tone lasts as long as the issuing station transmits the TOD signal.

### IFF

The IFF transmits coded selective identification to interrogating radar. The system operates in five modes and is capable of position identification and emergency identification. The modes provided are mode 1, mode 2, mode 3/A, mode 4, and mode C. Modes 1, 2, and 3/A are selectable.

Button L3 on UFC menu 1 controls code changes to Modes 1, 2, and 3, and can select the IFF menu. To change a code, enter mode number, dash, and two to four digits in the scratchpad. The UFC IFF identification key (I/P) initiates identification reply.

## RADIO NAVIGATION SYSTEM

The radio navigation system consists of a TACAN transceiver and antennas.

## TACAN

The TACAN provides slant range and bearing to a ground TACAN transmitting station, slant range to

## KY-58 MENU

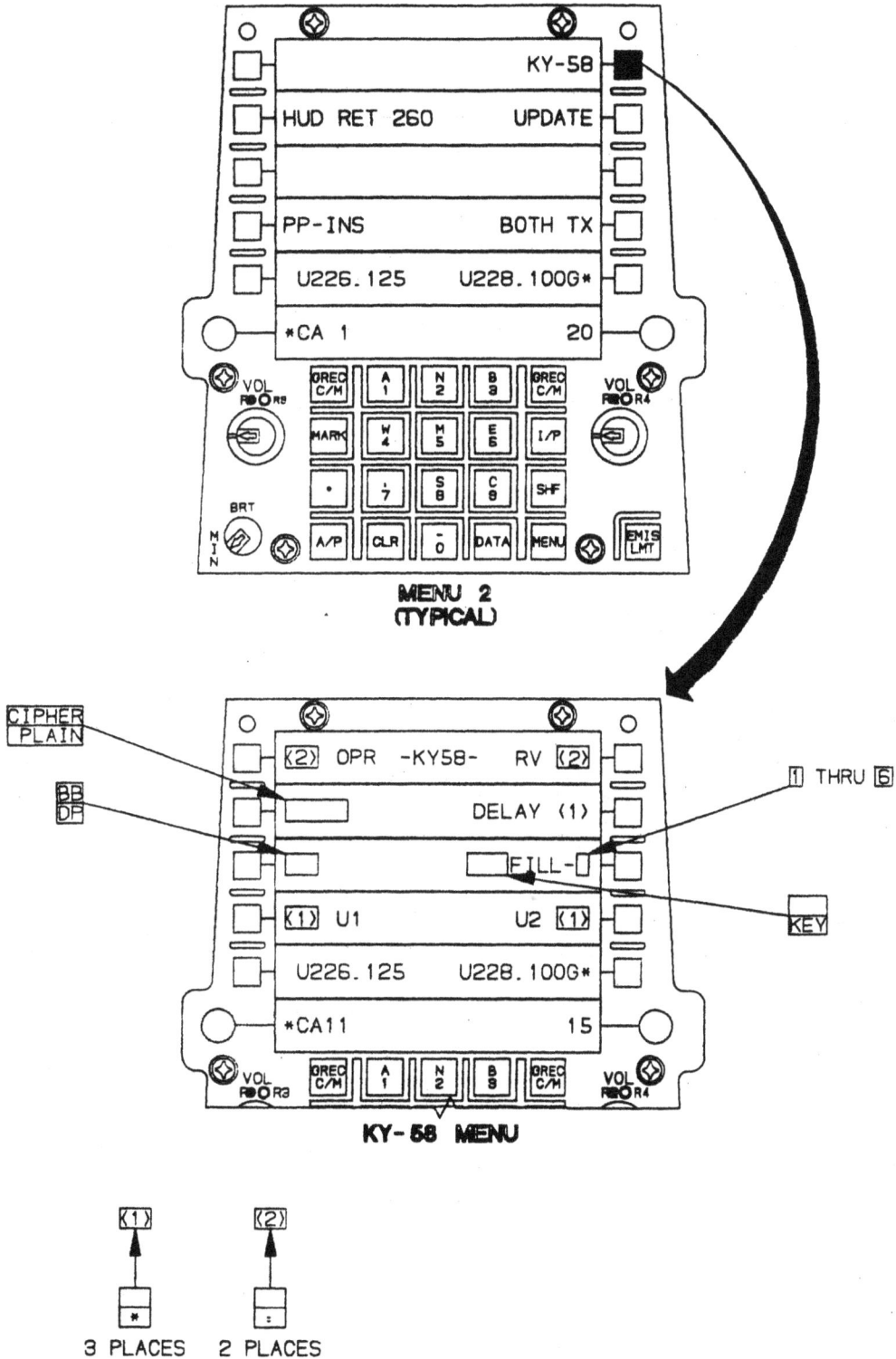

Figure 1-55

another aircraft with like equipment, and slant range/bearing to a suitably equipped aircraft. The TACAN interfaces with the avionics interface units and UFC for power control, operation, and station selection.

The 252 selectable channels are equally divided between X and Y channels. The 126 X channels are normally used for air-to-ground navigation. The 126 Y channels may be used in TACAN air-to-air option with cooperating aircraft minimizing interference with ground TACAN stations. Cooperating aircraft use channels exactly 63 channels apart.

### TACAN DISPLAY

The TACAN display (figure 1-56) is selected on the electronic HSI with the button labeled TCN. Navigation information is provided by a TACAN symbol, bearing pointer head and tail, course arrow, TACAN data block, the selected course readout, to/from symbol, heading marker, course deviation scale, and a course deviation indicator (CDI). Display range is adjusted by buttons at the top of the display.

The TACAN symbol is shown at the azimuth and distance from the aircraft symbol, proportional to the selected range. The symbol is shown when the TACAN station is within the display field of view. The symbol is not shown if the TACAN station is beyond the selected range of the display, the selected station is not operating, the system is unreliable, or TACAN data is invalid.

Other TACAN invalid indications include the removal of the course arrow, CDI, bearing pointer and to/from symbol. OFF also appears in the TACAN data block and above the boxed electronic HSI button labeled TCN.

The bearing pointer indicates bearing from the aircraft to the TACAN station. The bearing pointer points toward the TACAN if the symbol is not in the electronic HSI field of view and indicates the shortest turn direction to fly towards the TACAN.

The course arrow is oriented to the selected course readout. The course deviation scale appears either side of the course arrow. The selected course readout is adjusted by the arrow buttons above and below the readout. The heading marker (captain's bars) rotates around the compass rose and is adjusted by the arrow buttons above and below the heading marker readout.

The to/from symbol appears when TCN is selected. The station channel, radial, DME, and time to, or away from, the station appear in the TACAN data block.

### COURSE DEVIATION INDICATOR

The CDI moves within the compass rose of the TACAN display. On course is indicated when the CDI is aligned with the course arrow. Full scale deflection is 10 degrees.

### TACAN MENU

The TACAN menu provides control of TACAN channels and frequencies. Select the TACAN menu from UFC menu 1 by pressing L2 with no scratchpad entry (figure 1-57).

| L1<br>TCN | Applies (:) and removes TACAN system power. |
|---|---|
| R1<br>X | • TACAN frequency is X band. Steps to Y band. |
| Y | • TACAN frequency is Y band. Steps to X band. |
| L2<br>A/A | Selects (*) air-to-air TACAN. |
| R2<br>PROGRAM | Selects (*) TACAN program menu. Not used. |
| L3<br>T-R | Selects (*) TACAN transmit receive. |
| L4<br>REC | Selects (*) TACAN receive only. |

# INERTIAL NAVIGATION SYSTEM

The inertial navigation system (INS) is a nonemitting navigation, attitude, and heading reference system. Information is processed automatically and continuously. The INS consists of the INS control panel, ring laser gyros, accelerometers, and associated electronics.

The INS provides three-dimensional position, velocity, heading, magnetic variation, and attitude, and computes course deviation, wind direction and velocity, and distance and time to selected steer points.

### INS SELECTOR

#### INS Selector

The INS selector (figure 1-58) provides alignment options or inertial navigation selection. Rotating the selector to any position from OFF selects the present position menu on the UFC for INS coordinate entry.

## TACAN DISPLAY (TYPICAL)

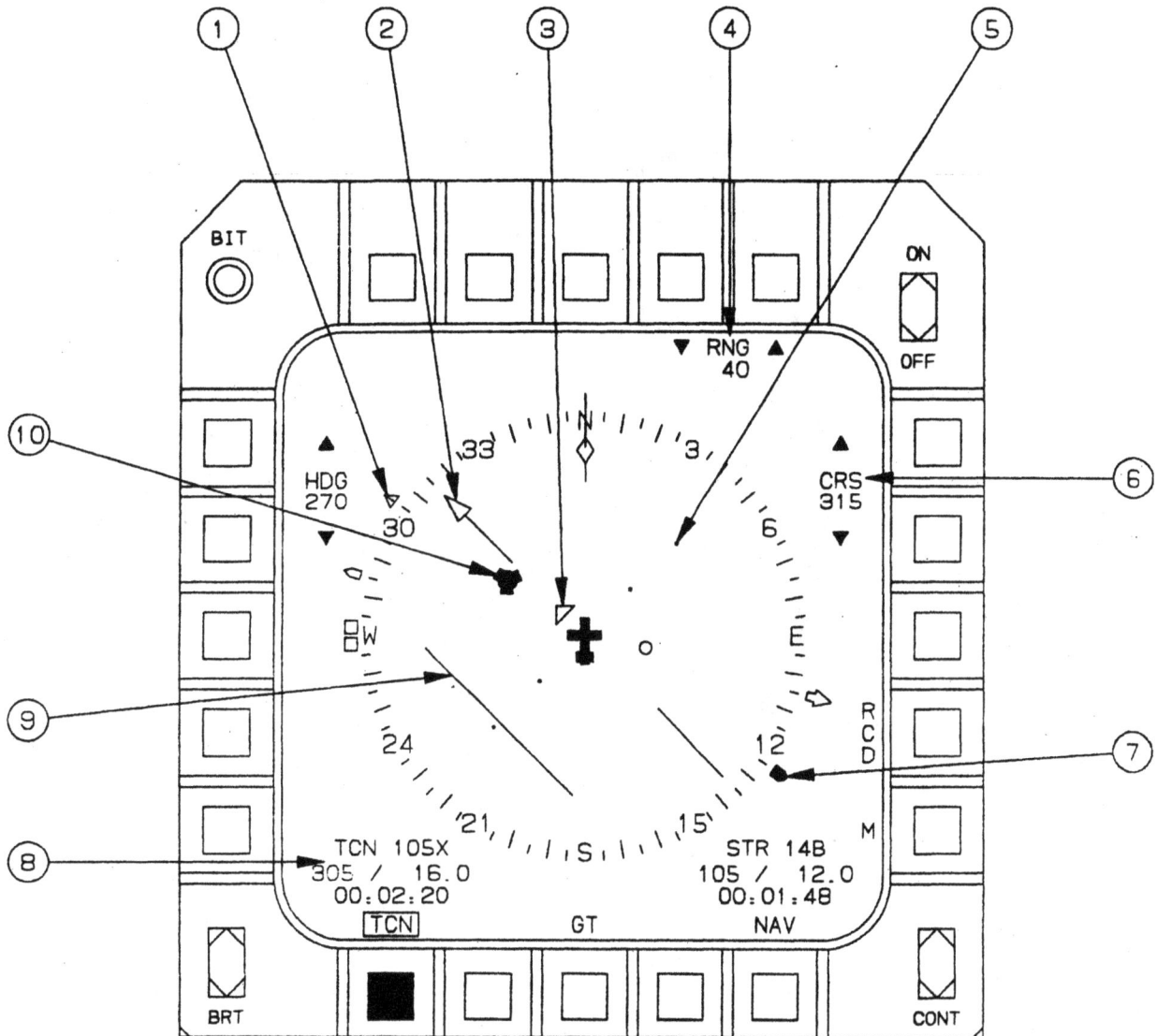

1. BEARING POINTER (HEAD)
2. COURSE ARROW
3. TO/FROM SYMBOL
4. DISPLAY RANGE
5. COURSE DEVIATION SCALE
6. SELECTED COURSE READOUT
7. BEARING POINTER (TAIL)
8. TACAN DATA BLOCK
9. COURSE DEVIATION INDICATOR
10. TACAN SYMBOL

A23-150

Figure 1-56

## TACAN MENU

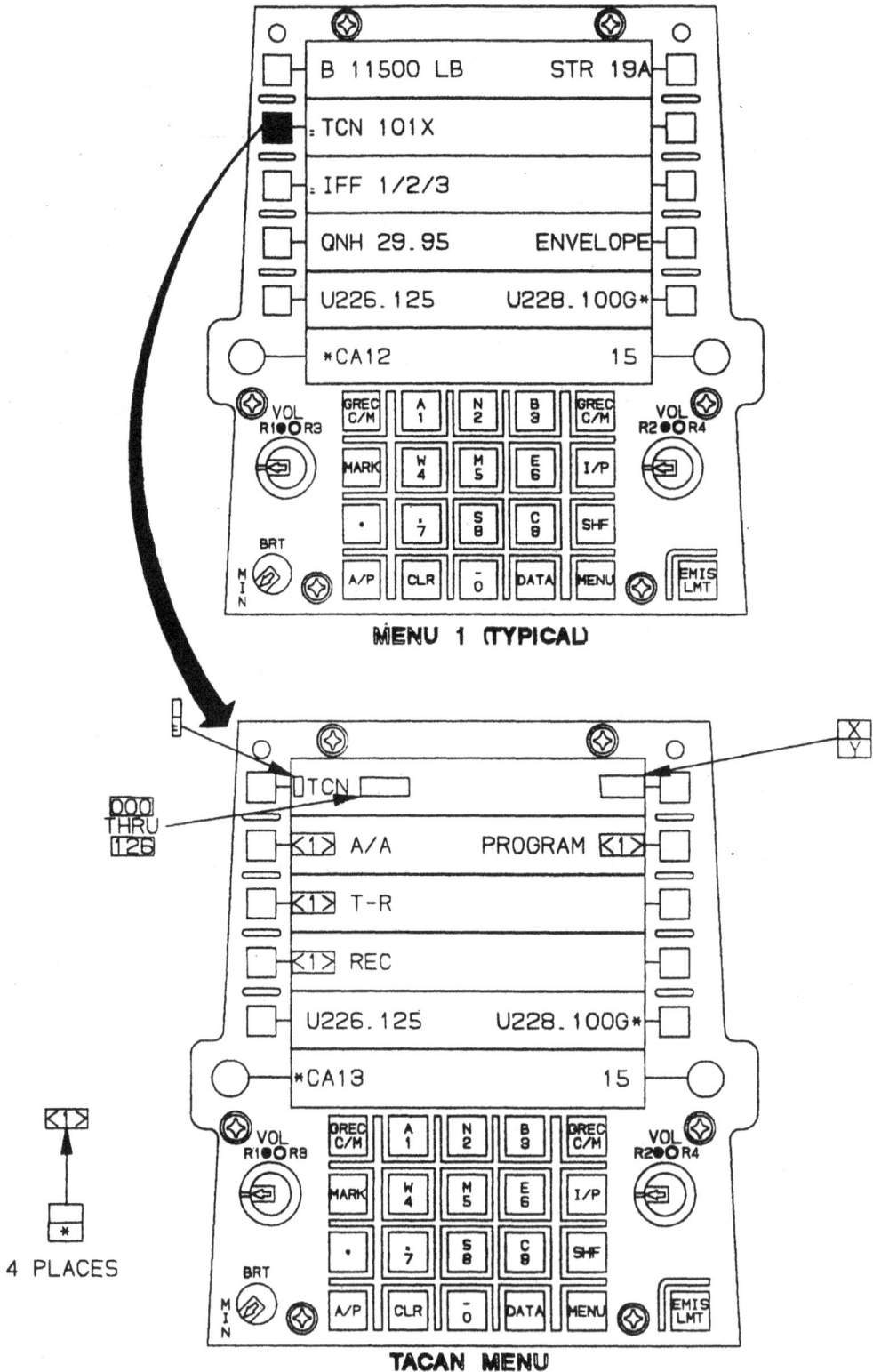

MENU 1 (TYPICAL)

TACAN MENU

4 PLACES

A29-151

*Figure 1-57*

## INS SELECTOR

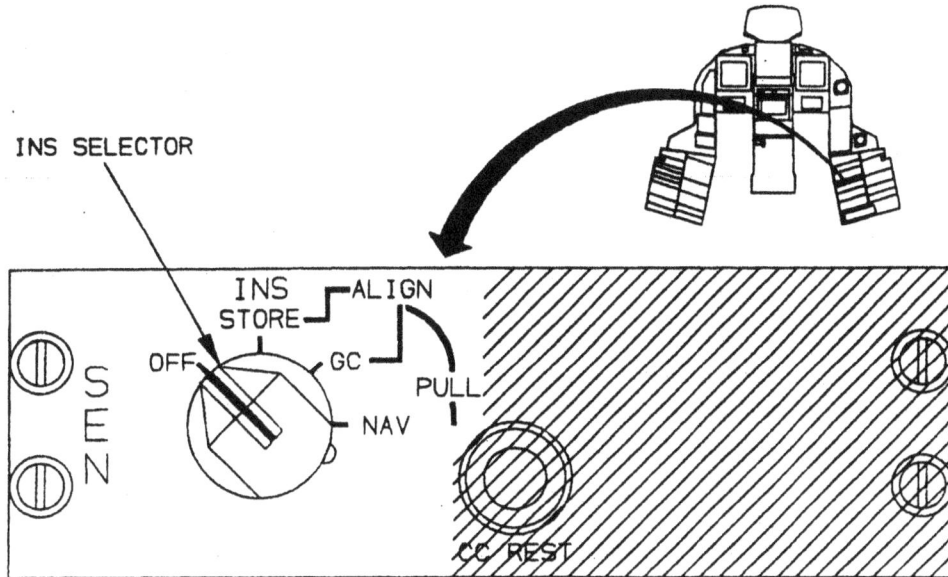

*Figure 1-58*

OFF      Power is removed from the INS.

STORE      Selects stored heading alignment using parameters stored during last system shutdown.

Stored heading alignment is achieved in approximately 40 seconds, providing the INS has completed a gyrocompass alignment without selecting NAV, powered down, and has not been moved. Accuracy depends upon the accuracy and error rate from the last system shutdown. If stored heading alignment is unavailable, the INS automatically selects gyrocompass alignment.

GC      Selects the gyrocompass alignment option.

Gyrocompass alignment requires approximately 4 minutes to complete.

NAV      Places INS into operation after alignment.

If selected immediately from OFF, the INS performs a gyrocompass alignment and transitions to NAV. To rotate out of NAV, pull selector out and counterclockwise.

### INS NAVIGATION DISPLAY

The INS navigation display (figure 1-59) is selected with the electronic HSI button labeled NAV. Inertial navigation information is provided by the command heading readout, navigation data block, and an INS bearing pointer head and tail.

Command heading readout shows the intercept direction to the INS bearing pointer head. The bearing pointer head indicates bearing from the aircraft to the steer point listed in the navigation data block.

### GROUND TRACK DISPLAY

The ground track display is selected with the electronic HSI button labeled GT. Ground track information is provided by the track marker and the selected course readout. The selected course readout presents the same information as the track marker heading. See Electronic HSI, this section, for illustration.

## TACTICAL SITUATION DISPLAY

The tactical situation display (TSD) provides an aircraft position relative to a planned mission route. Select the TSD with the TSD button on the MPCD menu display (figure 1-60).

## *INS NAVIGATION DISPLAY (TYPICAL)*

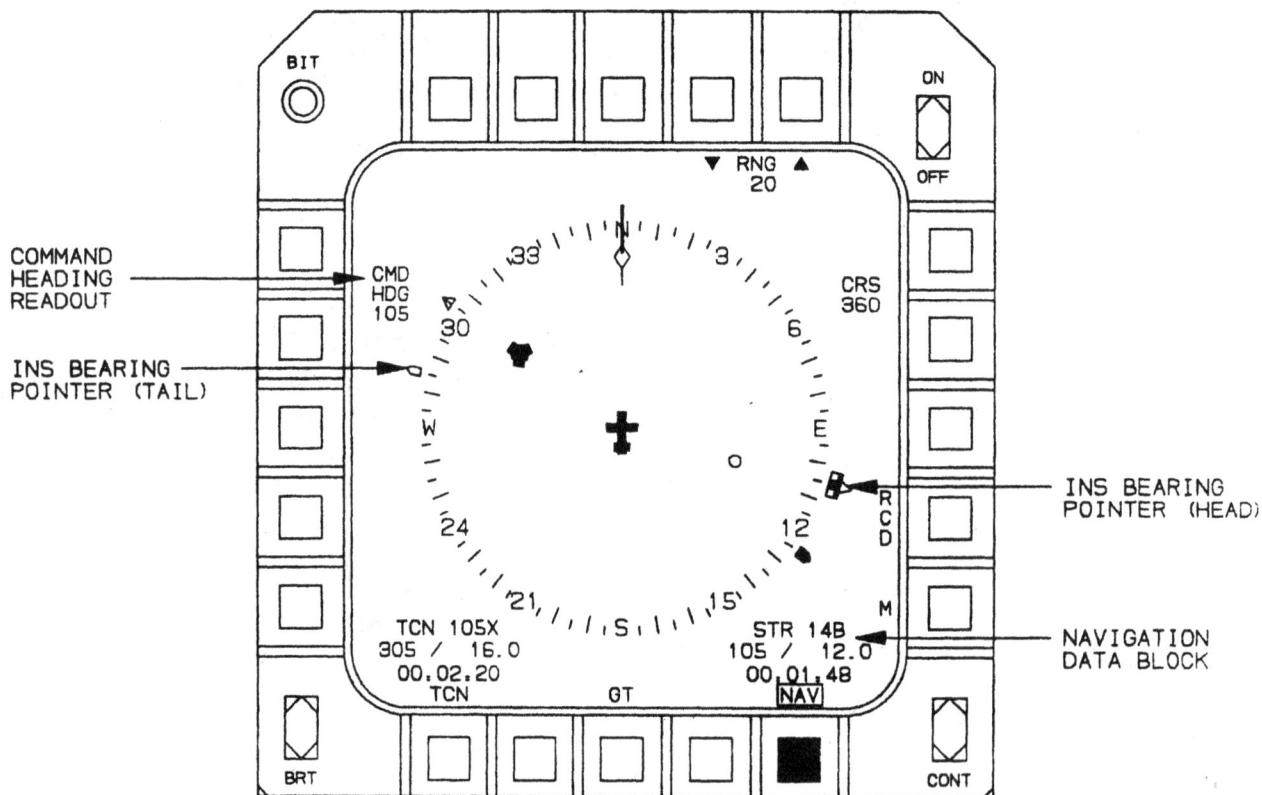

*Figure 1-59*

## TACTICAL SITUATION DISPLAY (TYPICAL)

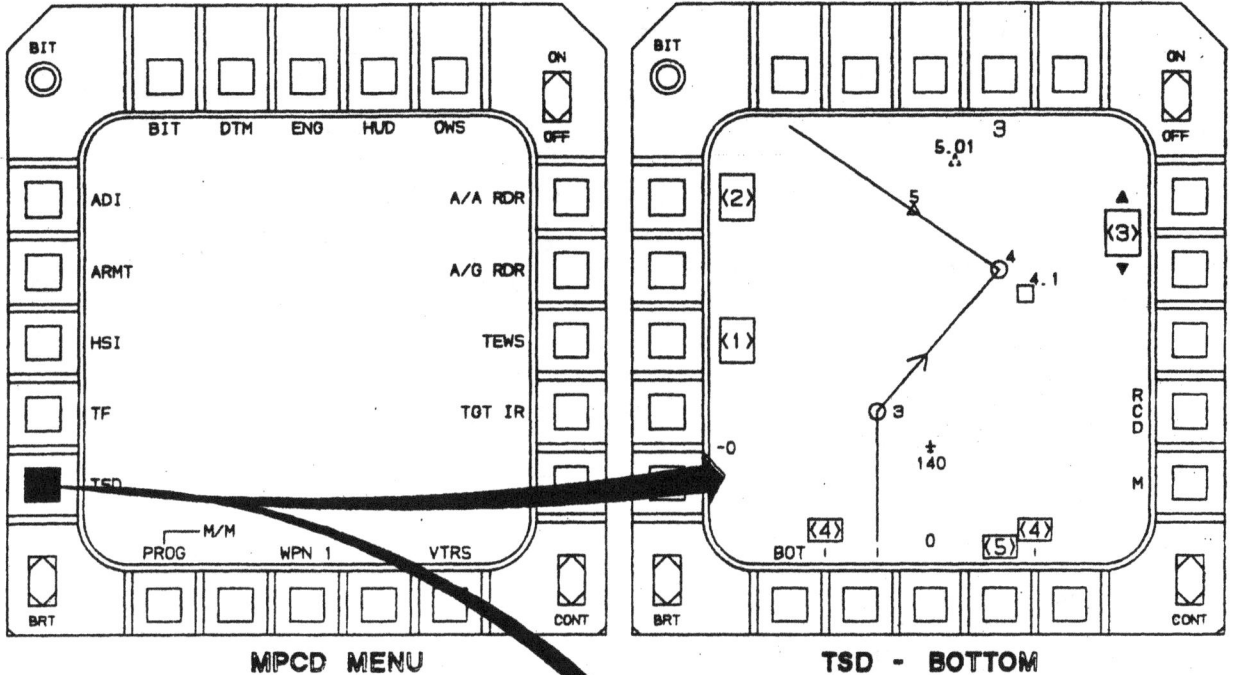

O   STEER POINT
□   AIM POINT
△   TARGET
⛬   OFFSET POINT
∨   MARK POINT
⌂   BASE
✤   AIRCRAFT POSITION
    WITH HEADING READOUT

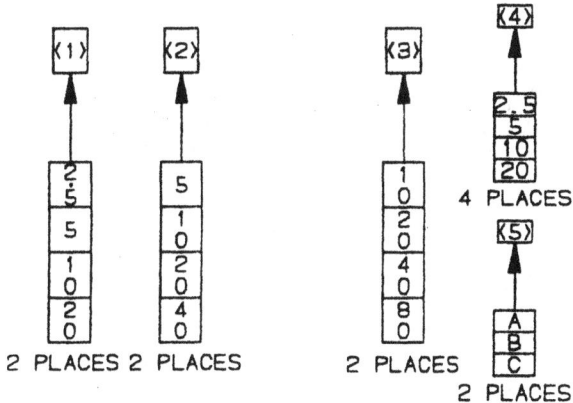

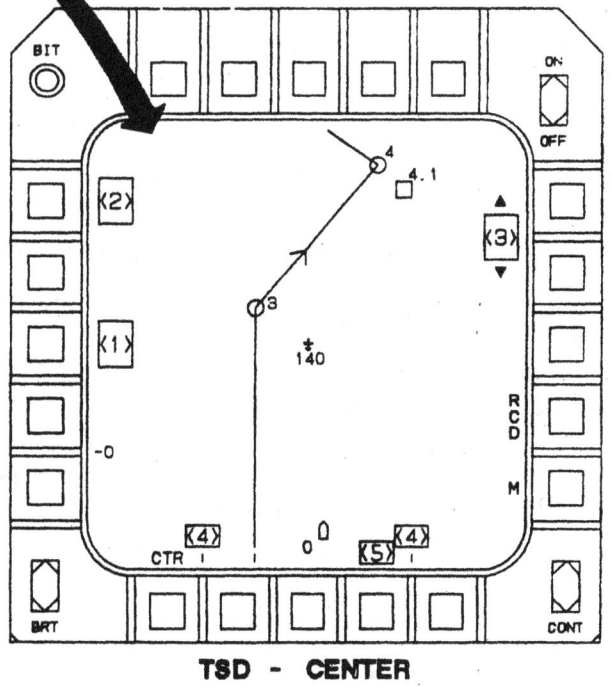

MPCD MENU

TSD - BOTTOM

TSD - CENTER

2 PLACES 2 PLACES    2 PLACES    4 PLACES    2 PLACES

A2∂-168

*Figure 1-60*

The selected scale is adjusted by the arrow buttons above and below the scale readout. Range tic marks on the display left side and bottom change proportionally as the selected scale is adjusted. Position can be estimated by viewing the aircraft symbol left or right of track, and judging relative distance from labeled sequence points. Inertial true course is presented below the aircraft symbol.

Two positions are available for the aircraft symbol. Selecting CTR centers the symbol on the TSD. Selecting BOT places the symbol on the lower 25 percent of the TSD.

The route identifier (A, B, or C), selected by the UFC, is presented on the lower right display. If an invalid sequence point is selected with the UFC, the advisory INVALID PP appears centered on the TSD. The display returns to the previous point selected in five seconds.

## UFC NAVIGATION CONTROLS

The UFC navigation controls direct INS, HUD, and TACAN data. The navigation controls can be selected from UFC menu 1, menu 2, or data 1.

## PRESENT POSITION MENU

The UFC present position menu shows latitude, longitude, and magnetic variation from INS, air data, or TACAN. The menu is used for initial entry of INS coordinates. The present position menu is selected by L4 on UFC menu 2 (figure 1-61).

Check the INS latitude and longitude shown are within 600 feet (0.100 arc minute) for the initial alignment. If not within 600 feet, enter the correct latitude and longitude into the scratchpad. Present position must be entered within 30 seconds after applying power to the INS, or the INS must be recycled. Subsequent position corrections are made with the update menu.

| | |
|---|---|
| L1 INS | Selects (*) present position from INS. |
| R1 A/D | Selects (*) present position from air data source. |
| R2 TCN | Selects (*) present position from TACAN. |
| L3 | Enters latitude. |
| R3 HUD TITL | Selects HUD titling menu. |

| | |
|---|---|
| L4 | Enters longitude. |
| R4 | Shows magnetic variation. |

## UPDATE MENU

The UFC update menu provides INS navigation updates from ground position steer point or a TACAN station. Select the update menu by R2 on UFC menu 2.

| | |
|---|---|
| R1 TCN | Selects (*) TACAN update. Shows TACAN channel. |
| L2 OFLY FRZ | Selects (*) overfly freeze update when directly over the steer point. |
| R2 X Y | Shows selected band TACAN frequency. |
| L3 NM | Shows difference north or south from the INS position to the update position. |
| R3 HUD | Not used. |
| L4 NM | Shows difference east or west from the INS position to the proposed update position. |
| R4 ENTER | Enters updated position into the INS. |

Ensure the steer point to be updated with the overfly freeze function is shown next to the legend UPDATE. steer point.

## STEER POINT MENU

The UFC steer point menu can show and change latitude, longitude, time-on-target, and minimum enroute altitude for labeled steer point. Select the steer point menu by R1 on either UFC menu 1 or data 1 (figure 1-62).

| | |
|---|---|
| L1 STR | Enters new steer point from scratchpad. |
| L2 | Enters steer point latitude from scratchpad. |

# PRESENT POSITION AND UPDATE MENUS

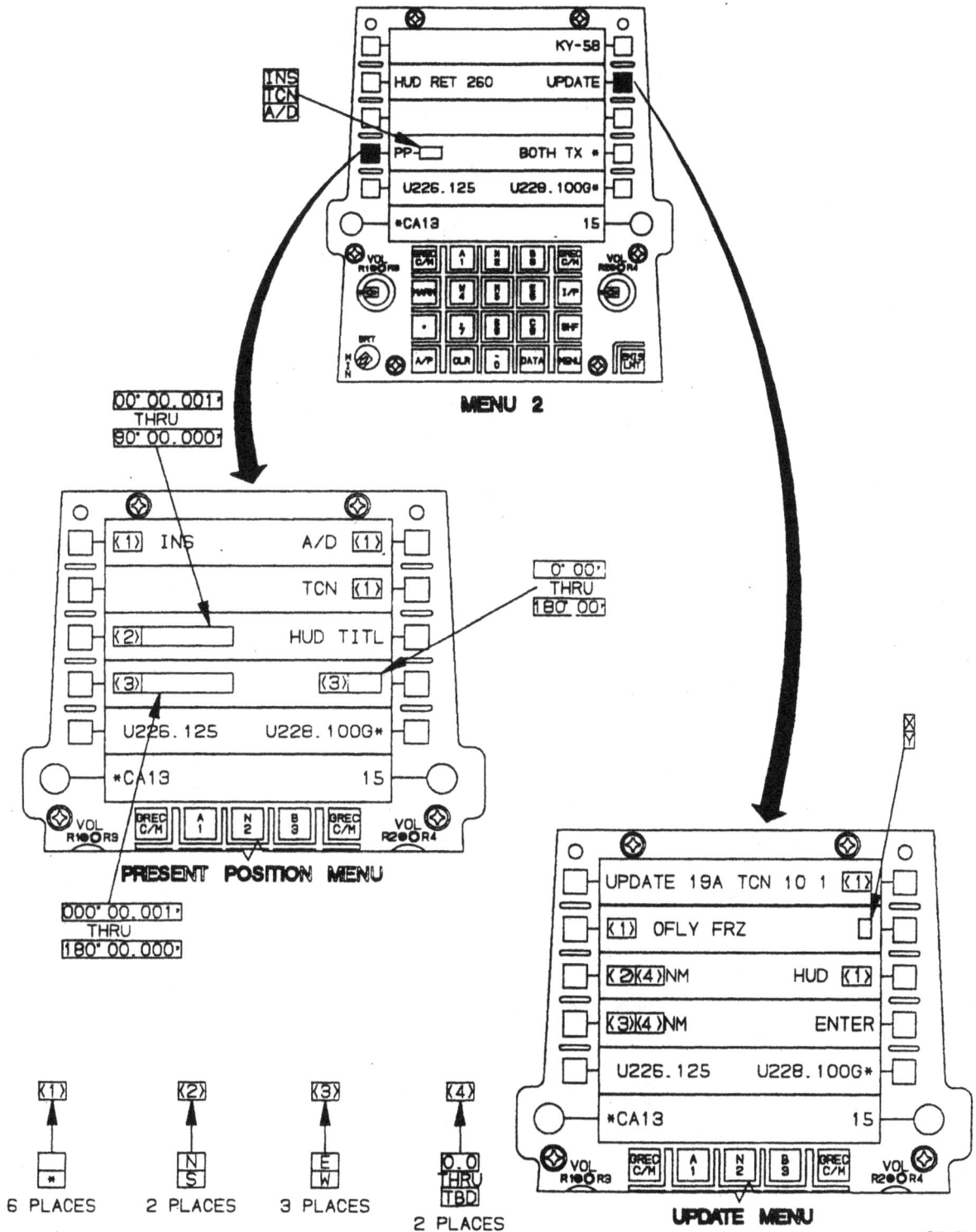

Figure 1-61

## STEER POINT MENU (TYPICAL)

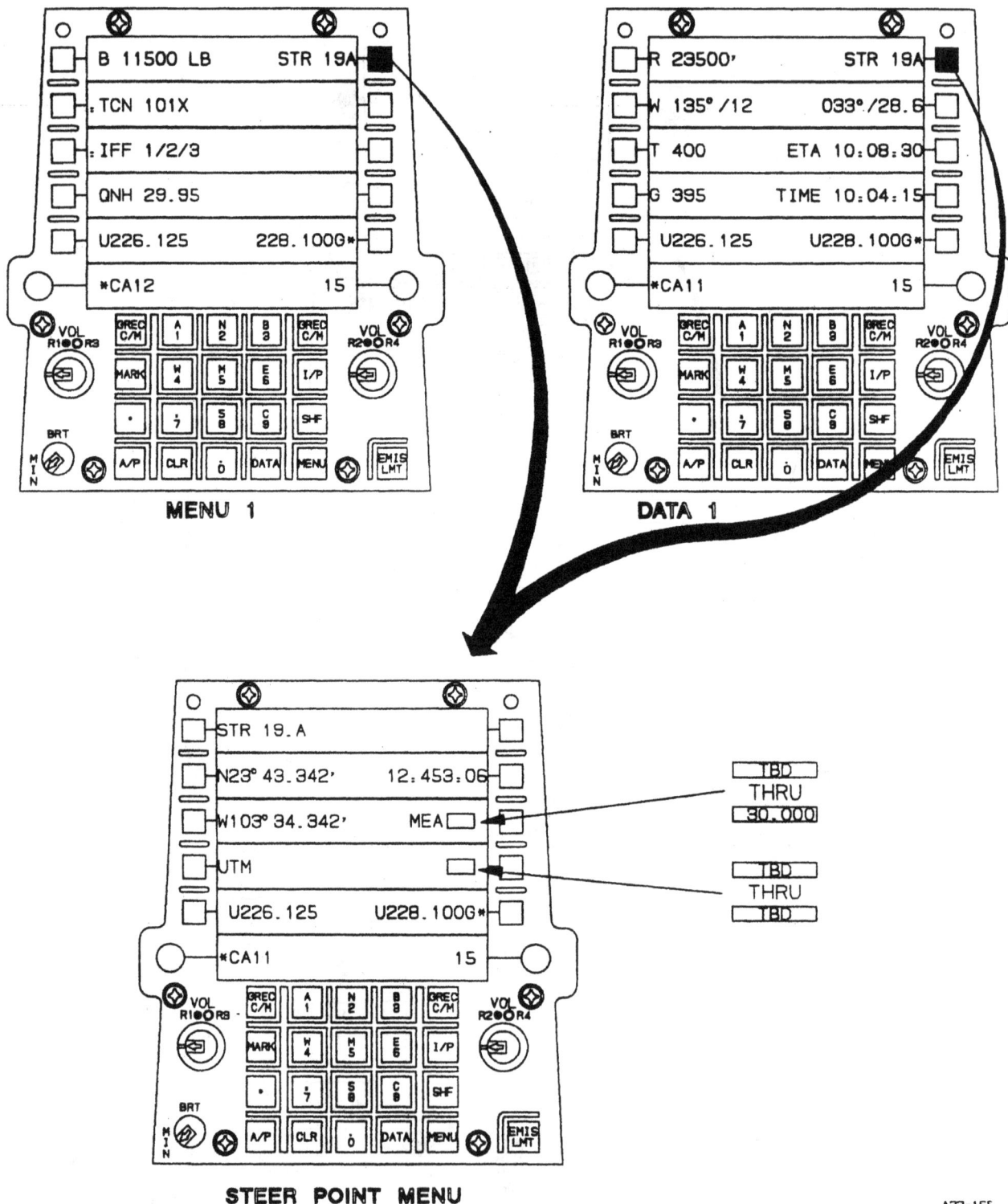

MENU 1

DATA 1

STEER POINT MENU

A23-165

*Figure 1-62*

R2
- Enters steer point time-on-target from scratchpad.
- Blank if no time-on-target has been entered.

L3

Enters steer point longitude from scratchpad.

R3
MEA

Enters steer point minimum enroute altitude from scratchpad.

L4
UTM

Selects the universal transverse mercator (UTM) menu.

R4

Enters steer point altitude from scratchpad.

## INS CAUTION

INS

- Inertial navigation system failure.
- Flight control laws in degraded mode.

# WARNING, CAUTION, AND ADVISORY SYSTEM

The warning, caution, and advisory (WCA) system provides an alerting system for hazardous (or potentially hazardous) conditions and changes in flight conditions requiring an awareness and possible action. The WCA system consists of red warning lights, aural warning **Tone** or *VOICE*, warning indicator lights, master caution light, color coded yellow caution and green advisory messages on the MPCD.

## WARNINGS

Red lights provide indication of system malfunction requiring immediate attention. Warning lights are located on the fire warning and controls panel, landing gear control panel, and warning lights panel.

### WARNING TONES AND VOICE WARNINGS

A warning **Tone** or *VOICE* accompany warning lights. The intercom panel warning tone silence button labeled V/W TONE SILENCE silences tones and voice warnings for 60 seconds. If the warning condition persists after 60 seconds, the tone or voice message is reactivated. A new warning signal activates a tone or voice warning without delay. The landing gear control panel warning tone silence button labeled WARN TONE SIL silences the warning tone associated with an unsafe landing gear condition.

## CAUTIONS

If a caution condition occurs, a caution message appears on the MPCD and the master caution (MC) light comes on.

### CAUTION MESSAGES

Caution messages normally appear on the right MPCD, but can be transferred and made to appear on the left MPCD by simultaneously pressing the master caution light and moving the control stick display selector (castle switch), left. Display information could be blanked to permit the cautions to be shown.

The cautions appear in order of occurrence. The most recent caution appears in the top right column. The word CAUTION appears on the display if AIU 1 (avionics interface unit) fails, indicating that most caution information is not being displayed.

### MASTER CAUTION LIGHT

The master caution light comes on when a caution message appears. An associated bell tone sounds twice for each master caution except the caution ENVELOPE has a wavering tone and voice message for the caution FUEL LOW. The master caution light goes out when the caution condition is cleared or the master caution light is pressed.

### MULTIPURPOSE DISPLAY EMERGENCY POWER

Caution messages are shown on the right MPCD if using emergency power or if central computer failure. The right MPCD is powered by the emergency generator. Cautions are shown on the bottom of the electronic ADI and can be removed by moving the control stick display selector. A second activation of the switch returns the cautions to the display.

## ADVISORIES

Advisories appear as either green or white lights, or green advisory messages on the MPCD.

The advisory BINGO activates the voice warning *BINGO FUEL*.

## WARNING AND ADVISORY LIGHTS

The interior light test switch, tests warnings and advisory lights.

The light indications are:

- LH/RH OVHT,
- OXY,
- MASTER CAUTION,
- LANDING GEAR WARNING,
- LEFT, NOSE, RIGHT LANDING GEAR INDICATORS,
- L/R ENG/AMAD FIRE,
- APU/EPU FIRE,
- IFF REPLY,
- APU READY,
- RCD, EDT,

# LIGHTING SYSTEM

The aircraft lights support daytime operations.

## EXTERIOR LIGHTS

Exterior lights consist of two anticollision strobe lights. The exterior lights are controlled from the lights panel (figure 1-63).

### Anticollision Light Switch

The three position exterior anticollision light toggle switch controls the intensity of the anticollision lights.

ON          Both anticollision lights flash bright.

OFF          Both anticollision lights off.

DIM          Both anticollision lights flash dim.

## UTILITY LIGHT

The utility light is located on the right console and operates from the essential dc bus.

### INTERIOR LIGHT TEST SWITCH

The two position, spring loaded off, interior light test switch tests interior warning, caution, and advisory lights. The lights remain on as long as the switch is held on.

## *LIGHTS AND CONTROLS*

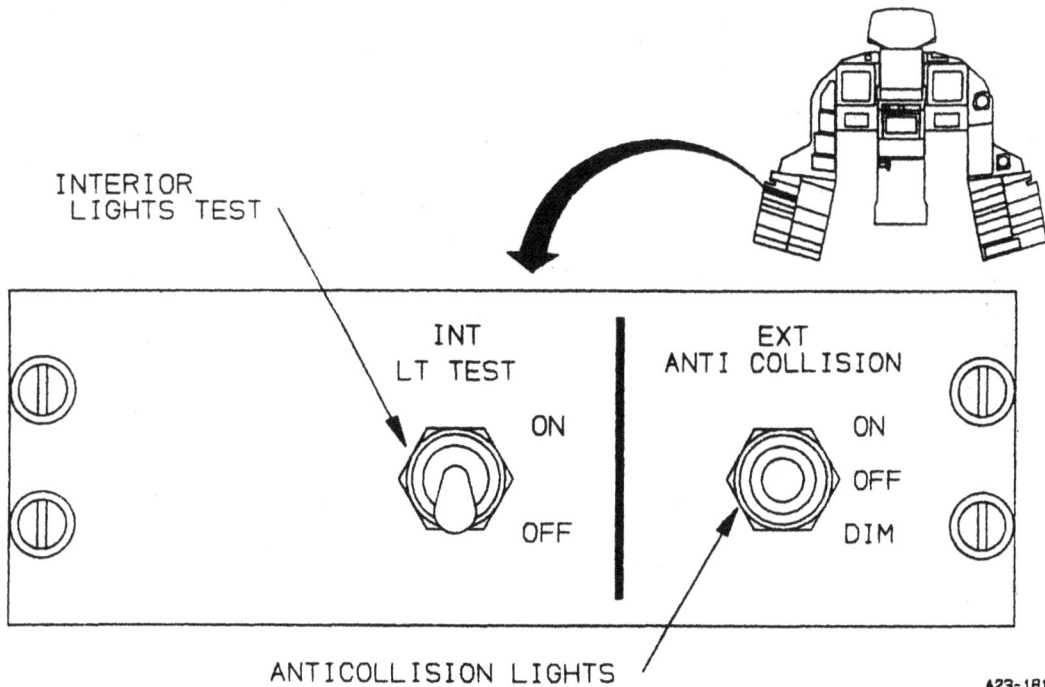

INTERIOR
LIGHTS TEST

INT
LT TEST

ON

OFF

EXT
ANTI COLLISION

ON

OFF

DIM

ANTICOLLISION LIGHTS

A23-181

**Figure 1-63**

# LIFE SUPPORT SYSTEMS

Life support systems include oxygen generation, anti-G protection, and personal cooling. Oxygen, anti-G protection, and personal temperature control are provided when used with a positive pressure breathing oxygen mask, helmet with mask tightening bladder, a torso counterpressure garment (jerkin), and the standard anti-G suit. Below 43,000 feet MSL, oxygen and anti-G protection can be provided with standard personal equipment.

Positive pressure breathing increases the ability to resist the effects of altitude and G exposure. Pressure in the the jerkin, mask, helmet bladder, and anti-G suit increase with G and altitude. The jerkin applies chest pressure to aid breathing. The helmet bladder tightens the oxygen mask to eliminate leaks. The jerkin and the helmet bladder are liquid cooled to reduce thermal stress.

## OXYGEN SYSTEM

The oxygen system consists of an oxygen concentrator, oxygen monitor, and electronic oxygen regulator that provide a continuous supply of oxygen enriched air when the ECS is operating. Conditioned engine bleed air and electrical power are supplied to the concentrator whenever the ECS panel mode switch is not in OFF/RAM and the ECS panel cabin pressure switch is not in RAM/DUMP.

The ac powered oxygen concentrator removes water vapor and nitrogen from the bleed air. The oxygen monitor continuously samples the concentrator output for oxygen concentration and provides a warning if concentration is insufficient. The oxygen regulator receives a dilution control signal from the monitor and automatically provides the correct oxygen concentration and pressure based on altitude, G, and oxygen regulator switch positions. The oxygen rich mixture is routed through the personal equipment connections to the oxygen mask, helmet bladder, and jerkin.

## OXYGEN REGULATOR

The electronic oxygen regulator dilutes the oxygen concentration with air based on the oxygen monitor concentration signal. Oxygen concentration increases to maximum by 29,000 feet cockpit pressure. Between 29,000 and 39,000 feet, oxygen is at a constant safety pressure. From 39,000 to 60,000 feet pressure increases linearly for altitude positive pressure breathing. G-biased positive pressure breathing occurs between 3.3 and 9.0 positive G.

The oxygen regulator is powered by the essential dc bus and controlled by switches on the regulator panel. The oxygen regulator panel has a maintenance test port, a flow indicator, and four switches (figure 1-64).

### Flow Indicator

The flow indicator light comes on with an oxygen flow through the regulator. At low oxygen concentrations, a deep breath could be required to indicate a flow of oxygen. The indicator light remains off if the power switch is off.

### Jerkin Switch

The two position jerkin switch is used if a jerkin is worn. The switch controls the pressure supplied to the mask and the jerkin. When subjected to G forces, the electronic anti-G valve sends a signal to the electronic oxygen regulator. If the jerkin switch is in the position ON, the regulator increases the pressure to the jerkin and the oxygen mask for G biased pressure breathing.

To prevent uncomfortable high pressure breathing pressures, the jerkin switch should be in OFF if the jerkin is not worn.

| | |
|---|---|
| ON | Oxygen enriched air at G biased pressure is delivered to the oxygen mask and jerkin. |
| OFF | Oxygen enriched air without G biased pressure is provided to the oxygen mask only. |

### Oxygen Supply Switch

The two position oxygen supply switch controls the flow of oxygen enriched air to the oxygen regulator. If emergency oxygen is activated, the switch should be turned off.

| | |
|---|---|
| ON | Oxygen rich mixture is supplied to the oxygen regulator. |
| OFF | Flow turned off and regulator does not operate. |

### Oxygen Switch

The three position oxygen switch controls oxygen enriched air provided to the oxygen mask.

| | |
|---|---|
| EMERG | Maximum oxygen concentration is supplied at positive pressure. |

## OXYGEN REGULATOR

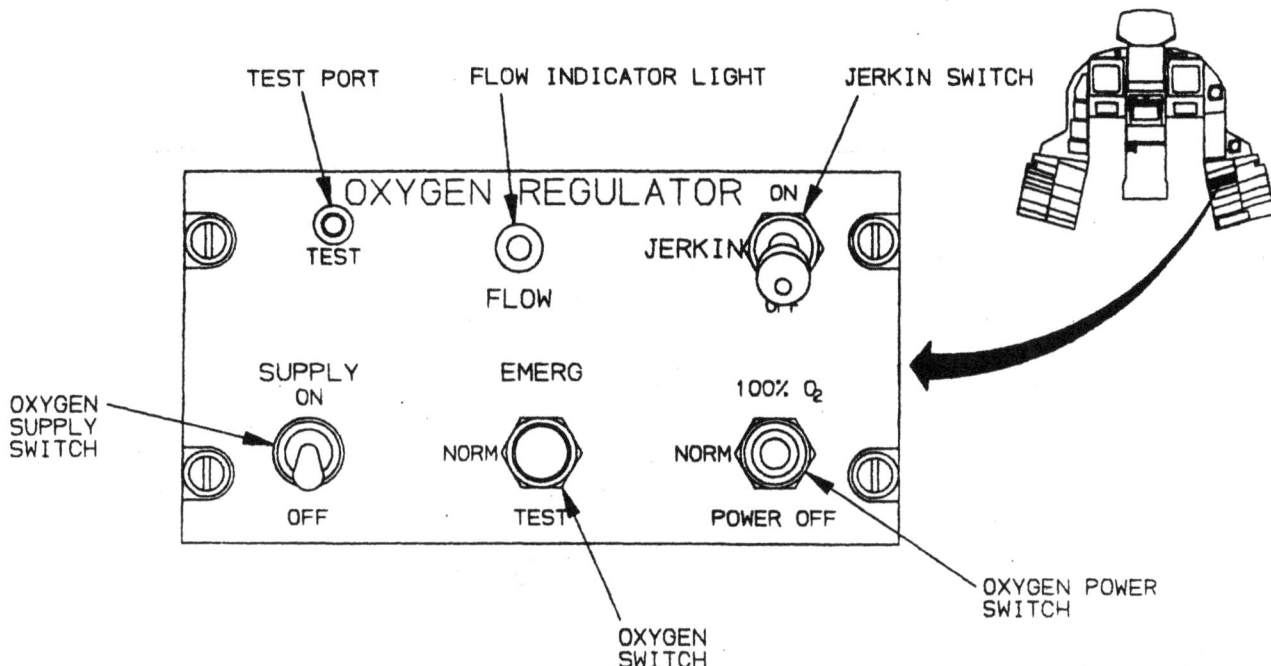

**Figure 1-64**

| | | |
|---|---|---|
| NORM | Oxygen enriched air is provided on demand. Oxygen concentration and pressure increase automatically with increasing cockpit altitude. | |
| TEST | Gradually increasing pressure is provided to test oxygen mask fit and hose connections. | |

regulator operates in the electronic mode. If jerkin switch is ON, oxygen enriched air at G biased pressure is provided.

100% — Maximum oxygen enrichment is provided at a slight positive pressure.

**Oxygen Power Switch**

The three position oxygen power switch controls electrical power to the oxygen regulator. If emergency oxygen is activated, the switch should be turned off.

| | |
|---|---|
| POWER OFF | Electrical power to the oxygen regulator is turned off. The regulator operates in a mechanical mode and provides oxygen enriched air without G biased pressure. |
| NORM | Electrical power to the oxygen regulator is turned on and the |

**OXYGEN MONITOR**

The oxygen monitor (figure 1-65) monitors oxygen concentration. The oxygen monitor is to the right of the ejection seat under the canopy sill. The oxygen monitor has a test button to check the oxygen monitor, oxygen concentrator, and warning circuitry.

**LIFE SUPPORT CONNECTIONS**

Life support connections consist of the torso harness connected breathing manifold, the life support connections panel, and the anti-G valve. The breathing manifold provides normal and emergency oxygen hose connections. The oxygen mask, helmet bladder/skull cap, and jerkin connect to the breathing manifold. The life support connections panel

## *LIFE SUPPORT CONNECTIONS*

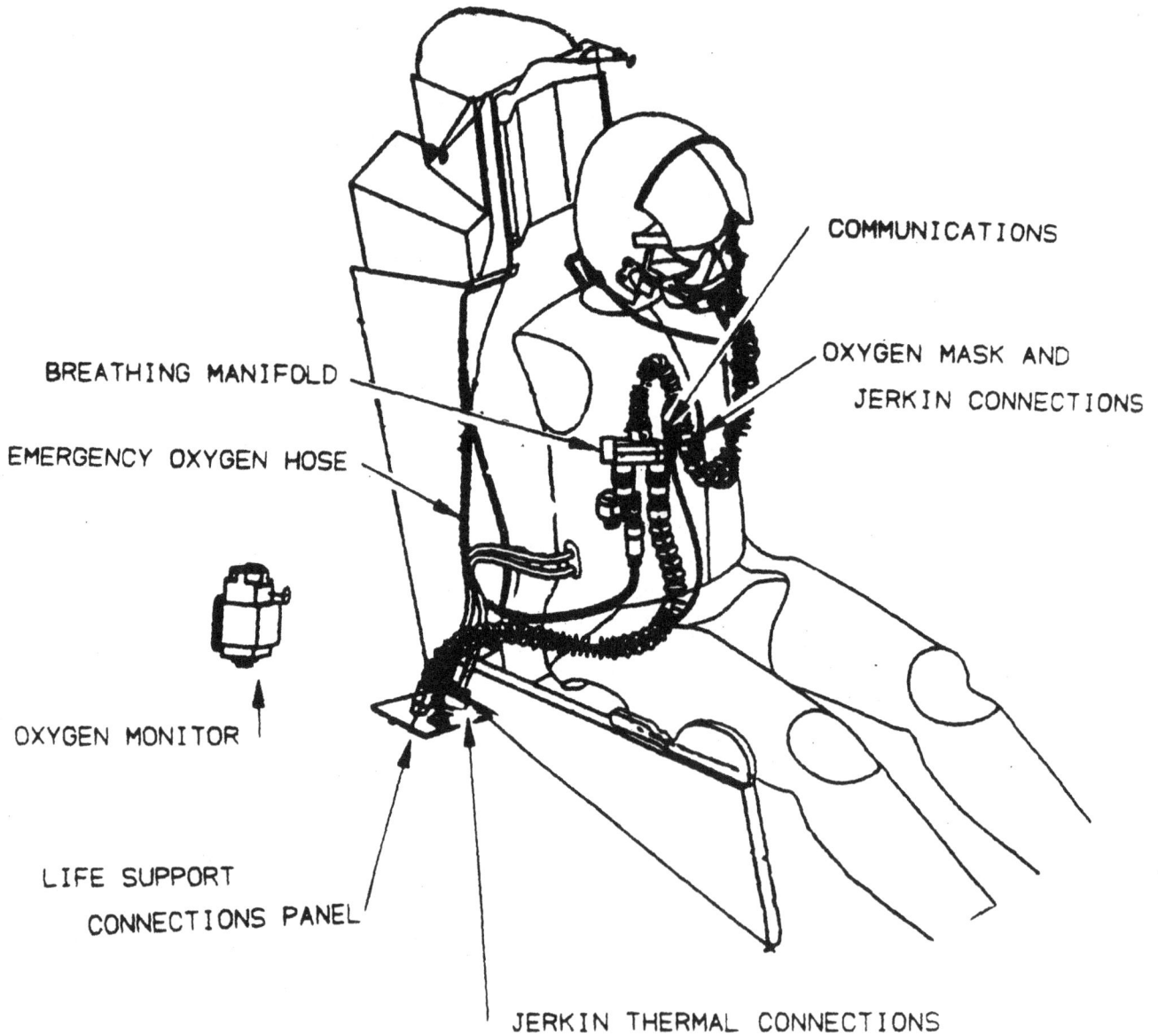

COMMUNICATIONS

OXYGEN MASK AND
JERKIN CONNECTIONS

BREATHING MANIFOLD

EMERGENCY OXYGEN HOSE

OXYGEN MONITOR

LIFE SUPPORT
CONNECTIONS PANEL

JERKIN THERMAL CONNECTIONS

Figure 1-65

provides connections for the communications, oxygen, and jerkin thermal connections. The G-suit is connected to the anti-G valve. See figure 1-66.

### Jerkin Thermal Control

A thermal control unit provides temperature control for liquid cooling of the jerkin and skull cap through the life support connections panel. When rotated either direction out of OFF, the temperature selector turns the thermal control unit on and controls the flow rate of the coolant. The position COOL is maximum cooling and flow rate.

Indication of system failure is shown by two lighted legends on the panel. System should be turned off and attempt to reset.

| | |
|---|---|
| NO FLOW | Coolant circulation failure. |
| INOP | Coolant refrigeration has failed. |

### Anti-G Valve

The electronic anti-G valve is powered by the essential dc bus and uses pressure from the ECS for anti-G suit inflation. The valve provides increasing pressure to the anti-G suit as normal G increases. When enabled through the oxygen regulator jerkin switch, the G suit inflates for positive pressure breathing. The anti-G valve has a two position lever lock switch, spring loaded out of TEST, and a G suit connection.

| | |
|---|---|
| TEST | An increasing test pressure is applied to the G suit. If a jerkin is worn and the oxygen regulator jerkin switch is in ON, pressure is applied to the jerkin. |
| NORMAL | Pressure is proportional to G forces. When positive pressure breathing is enabled, pressure is four times the pressure provided to the oxygen mask. |

### OXYGEN WARNING

| | |
|---|---|
| OXY *THREAT WARNING*. | • Oxygen concentration low.<br>• Oxygen generator unit failed.<br>• Oxygen regulator failed. |

## PERSONAL EQUIPMENT CONTROLS

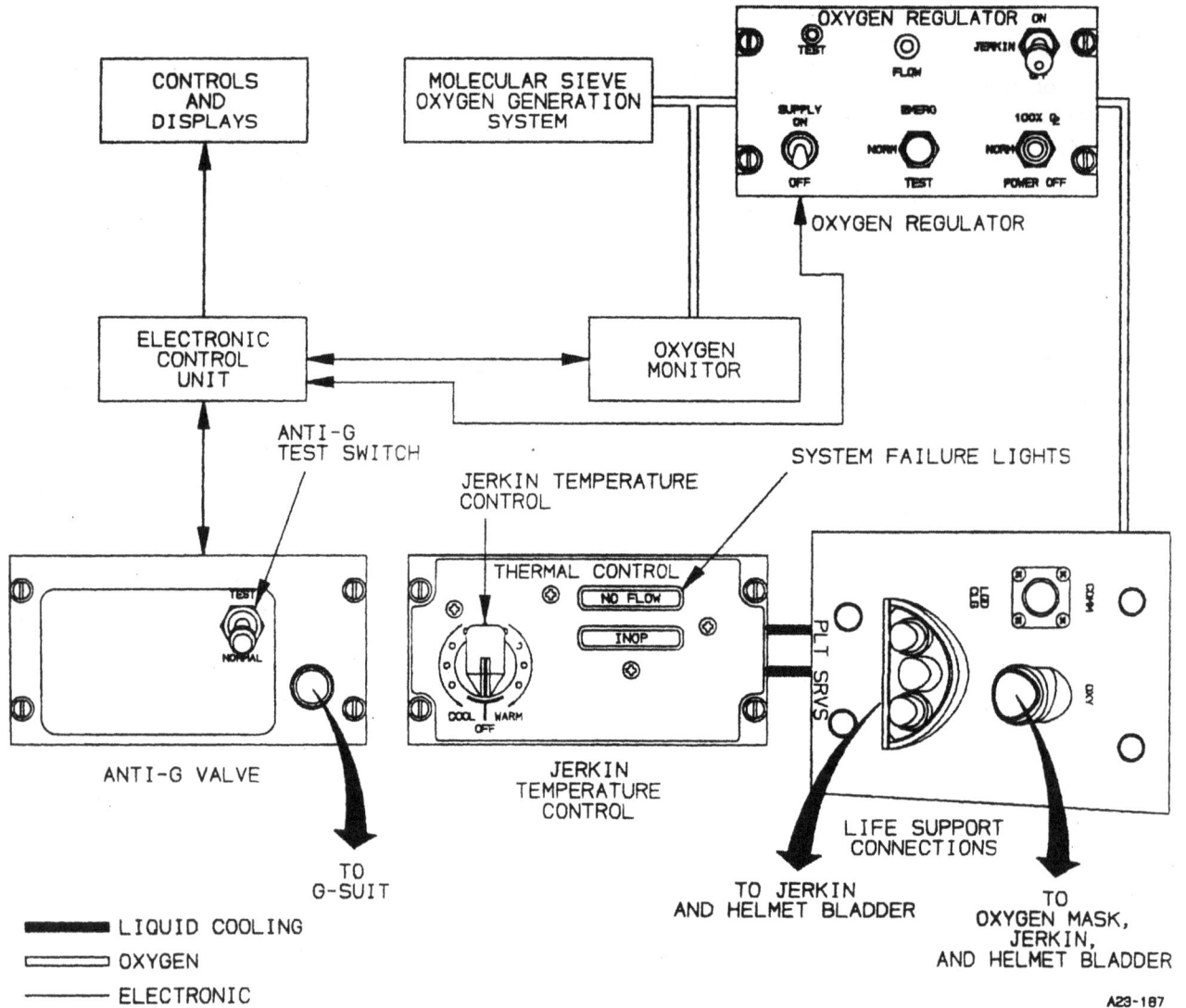

*Figure 1-66*

# EJECTION SEAT SYSTEM

The advanced concept ejection seat (ACES II) is a catapult rocket ejection system. The ejection seat provides escape in the 0 to 600 knots ejection envelope. Controls are provided to adjust seat height and lock the shoulder harness. Survival kit, radio beacon transmitter, and emergency oxygen supply are incorporated into the ejection seat, figure 1-67.

## SEAT CONTROLS

### Seat Adjust Switch

The seat adjust switch on the left console is a three position spring-loaded to center switch. The adjust switch must be released when the seat reaches an upper or lower limit to prevent damage to the actuator motor.

### Inertia Reel

A dual strap shoulder harness inertia reel is mounted in the seat below the head rest pads. The inertia reel automatically locks when rapid rate of strap reel out is sensed. Manual locking/unlocking of the reel is controlled by the inertia reel lock handle.

### Inertia Reel Lock Handle

The spring tensioned inertia reel lock handle has two positions.

| LOCKED | • The inertia reel straps prevented from being extended.<br>• Slack in the straps is pulled back into the reel. |
|---|---|
| UNLOCKED | Unlocks inertia reel. |

## EJECTION MODES

The ejection seat automatically selects one of the three ejection modes Pitot tubes on the head rest snap up on ejection to measure the ejection altitude and airspeed.

### Mode 1

Mode 1 is a slow speed/low altitude mode. The parachute is deployed immediately after the seat departs the aircraft.

### Mode 2

Mode 2 is a high speed/low altitude mode. The seat drogue chute deploys and slows the seat before man-seat separation. The parachute deploys after man-seat separation.

### Mode 3

Mode 3 is high speed/high altitude mode. Man-seat separation and parachute deployment are delayed until mode 2 or mode 1 parameters are met.

## EJECTION CONTROLS

### Ejection Handles

An ejection handle is on either side of the seat. The handles are interconnected so actuation of either handle initiates ejection.

### Ejection Seat Safety Lever

The ejection seat safety lever is immediately aft of the left ejection handle. With the safety lever rotated up and forward, both ejection control handles are mechanically locked.

### Restraint Emergency Release Handle

The restraint emergency release handle on the right arm rest is a backup to automatic man-seat separation. The handle releases the lap belt, the inertia reel straps, the seat pan, and the parachute pilot chute. The trigger on the inside of the handle grip is squeezed and then the handle is pulled vertically until the reaching full travel.

### WARNING

Do not pull the handle in flight. Pulling the handle disconnects the lap belt and inertia reel straps, making safe ejection impossible.

## SURVIVAL KIT

The survival kit consists of a fabric case housing a life raft, rucksack, and an auxiliary container. The life raft and rucksack are attached to the survival kit case by a drop line. The auxiliary container, used for storage of items to be retained, is secured inside the survival kit. An AN/URT-33C radio beacon is installed in the kit. The survival kit stows in the seat bucket beneath the rigid seat pan. The pan pivots for withdrawal of the kit during man-seat separation. The kit is attached to the torso harness by attachment fittings on the kit retaining straps. A survival kit auto/manual deployment selector permits either manual or automatic deployment of the kit.

### Survival Kit Deployment Selector Window

| A<br>(Forward) | The kit automatically deploys after man-seat separation. |
|---|---|

## EJECTION SEAT

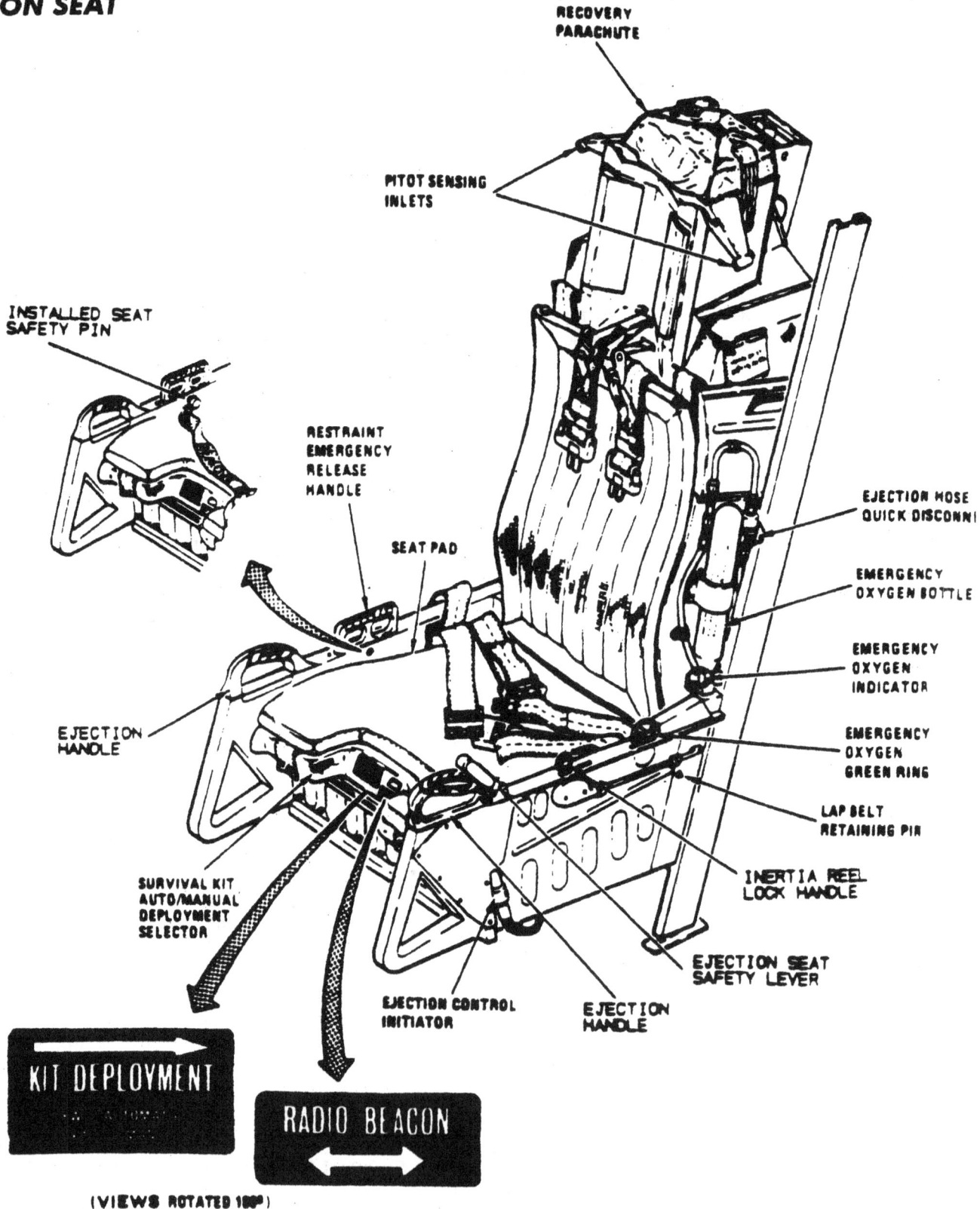

RECOVERY PARACHUTE

PITOT SENSING INLETS

INSTALLED SEAT SAFETY PIN

RESTRAINT EMERGENCY RELEASE HANDLE

SEAT PAD

EJECTION HOSE QUICK DISCONN!

EMERGENCY OXYGEN BOTTLE

EMERGENCY OXYGEN INDICATOR

EMERGENCY OXYGEN GREEN RING

LAP BELT RETAINING PIN

EJECTION HANDLE

INERTIA REEL LOCK HANDLE

SURVIVAL KIT AUTO/MANUAL DEPLOYMENT SELECTOR

EJECTION SEAT SAFETY LEVER

EJECTION CONTROL INITIATOR

EJECTION HANDLE

KIT DEPLOYMENT

RADIO BEACON

(VIEWS ROTATED 180°)

**Figure 1-67**

**NORTHROP/MCAIR YF-23A**
**COMPETITION SENSITIVE**
**UNCLASSIFIED**

| M (Aft) | After man-seat separation, the survival kit must be manually deployed by the survival kit release handle. |
|---|---|

### AN/URT-33C RADIO BEACON

A cutout in the front of the seat pan provides access to the radio auto/manual rocker switch. Left selects MAN, right selects AUTO.

| MAN | Emergency radio does not activate at man-seat separation. |
|---|---|
| AUTO | Emergency radio activates at man-seat separation. |

### IFF EMERGENCY OPERATION

Upon ejection, the IFF emergency mode automatically becomes active if mode 1, 2, 3A, or C is enabled.

### EMERGENCY OXYGEN

An emergency oxygen bottle is visible for inspection. The oxygen supply hose is routed to the breathing manifold connector and is actuated automatically during an ejection. The oxygen can also be activated manually by a emergency oxygen green ring on the left side of the seat bucket. The normal pressure, 1800 psi, supplies approximately 5-15 minutes of oxygen.

### CANOPY SYSTEM

The cockpit is enclosed by an aft hinging clamshell type canopy. The canopy system consists of the canopy, canopy switches, an electrically driven rotary actuator for powered canopy operation, latches on the canopy frame to lock the canopy to the fuselage, and fuselage mounted locks. An explosive unlock mechanism and two canopy jettison rocket boosters provide emergency jettison. An inflatable seal seals the cockpit for pressurization when the canopy is closed and locked.

The canopy can be operated by a canopy switch in the cockpit or a switch in the nose wheel well. An external crank can be used to open the canopy if electrical power is unavailable.

### Canopy Switch

The three position, electro-magnetic, spring-loaded to center canopy switch raises the canopy when placed in the position labeled OPEN. The switch is electrically held until the canopy is full open and then the switch returns to center.

The center position, labeled HOLD, stops the canopy during an open or close cycle. The canopy can be stopped in any position.

The switch must be held in the position labeled CLOSE until the canopy is down and locked. When the canopy reaches the sill, the last 1.5 inches of actuator travel moves the canopy forward to lock the canopy. Lock is indicated by the master caution light going out and the canopy advisory being rescinded.

### External Canopy Switch

The three position, spring-loaded center external canopy switch on the left side of the nose wheel well is similar to the switch in the cockpit. The external switch must be held in the position OPEN until the canopy is fully open.

### Manual Canopy Crank

The back-up manual cranking mechanism is in the left avionics bay.

### Canopy Jettison

A cartridge-initiated thruster moves the canopy aft to unlatch the canopy. The canopy frame-mounted rocket motors rotate the canopy up and aft. Either ejection handle or the canopy jettison handle fires the thruster.

The canopy can be jettisonable from the closed, open, or any intermediate position.

A black and yellow striped canopy jettison handle with release button is on the left inboard canopy sill. The thumb detent unlock button, when pressed, permits the handle to be rotated and pulled to the mechanical stop for canopy jettison.

# FLIGHT TEST EQUIPMENT

The three major flight test-unique controls are the flight test function control panel, the flight test display panel, and the VMS test panel.

## FLIGHT TEST FUNCTION CONTROL PANEL

The flight test function control panel (figure 1-68) controls the collection of data, telemetry frequency, and operation of the tracking beacon.

### Flight Test Data Button

The alternating on/off flight test data pushbutton starts the flight test data magnetic tape recorder and the video tape recorder (if selected by the applicable camera control on the camera/air refueling panel). The legend ON comes on green.

### Data Mode Selector

The rotary data mode selector, labeled DATA MODE, turns on the data system if the ac and dc essential buses are powered.

| | |
|---|---|
| OFF | Removes electrical power from the data acquisition and recording system. |
| STANDBY | Provides electrical power to the data system for warm up. Electrical power is not connected to recording equipment. |
| A,B,C,D,E,F | Enables a predetermined mode of operation. |

### Telemetry Frequency Selector

The five-position telemetry frequency selector, labeled T/M FREQ, selects predetermined telemetry frequency 1, 2, 3, 4, or 5.

### Automatic Trim Record Button

The automatic trim record button, labeled ATR, starts a 14 second cycle to calibrate data.

### C-Band Transponder Switch

The two-position C-Band transponder toggle switch, labeled C BAND, turns the C-Band transponder on off.

### Event Tone Switch

The two-position tone toggle switch, labeled TONE, enables/disables the tone circuit for comm 1 radio.

## LIGHT TEST FUNCTION CONTROL PANEL

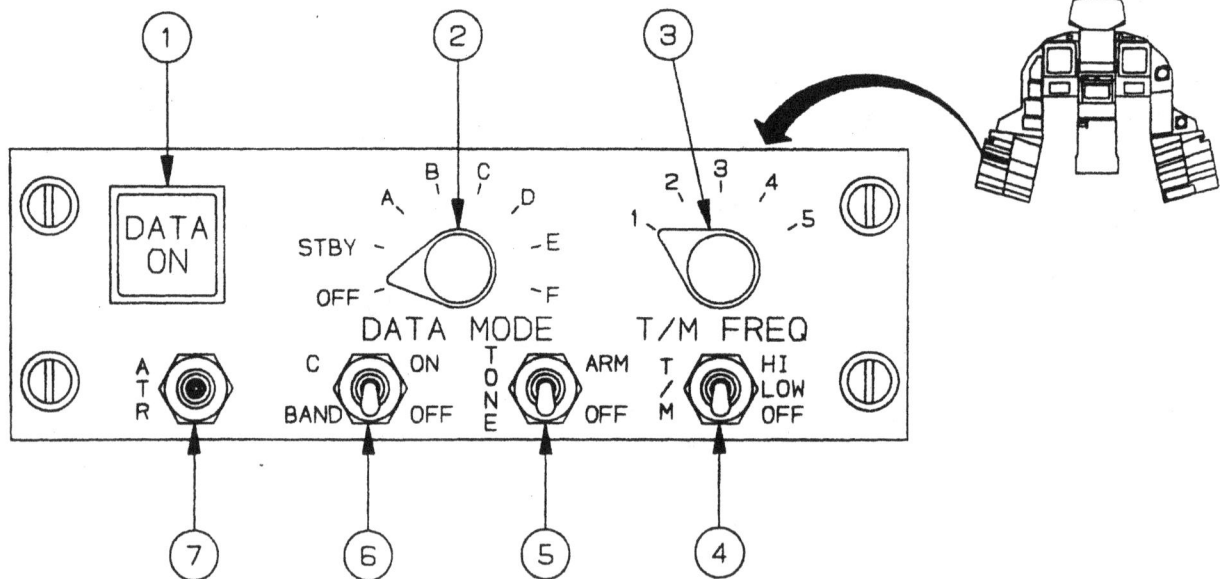

1. FLIGHT TEST DATA BUTTON
2. DATA MODE SELECTOR
3. TELEMETRY FREQUENCY SELECTOR
4. TELEMETRY POWER SWITCH
5. EVENT TONE SWITCH
6. C-BAND TRANSPONDER SWITCH
7. AUTOMATIC TRIM RECORD BUTTON

A28-210

*Figure 1-68*

ARM           Enables tone circuit. Pressing the control stick event button transmits a tone on the UHF radio.

OFF           Disables tone circuit.

### Telemetry Power Switch

The three-position telemetry power toggle switch, labeled T/M, selects the output power of the telemetry transmitter.

HI           Telemetry information is transmitted at 20 watts.

LO           Telemetry information is transmitted at 2 watts.

OFF           Telemetry transmitter is off

### FLIGHT TEST DISPLAY PANEL

The flight test display panel (figure 1-69) shows flight-test-unique information and has controls for the camera and data recording.

### Flight Test Data Pushbutton

The alternating on/off flight test data pushbutton with the legend DATA ON duplicates the flight test data switch on the flight test function control panel.

When either pushbutton is on, the legend ON comes on green on both pushbuttons.

### CAMERA CONTROLS

Either the weapons camera or the video camera is controlled by the camera/air refueling panel.

### Weapons Camera

The two position weapons camera switch starts or stops pre-selected weapons cameras. The top of fuselage camera records drag chute operation. The bottom of fuselage and under wing cameras record weapons separation.

### Video Cameras

The three position video camera toggle switch controls two hours of video recording of the MPCD/HUD or the over the shoulder video. If hot mic is selected, voice is recorded.

The position MPCD provides for split screen recording of both MPCD, an MPCD and the HUD, or the HUD only. When a display on the MPCD has a record function (RCD) and the function is selected, the display is split screen with another display or with the HUD. The label RCD is replaced with the time

## *FLIGHT TEST DISPLAY PANEL*

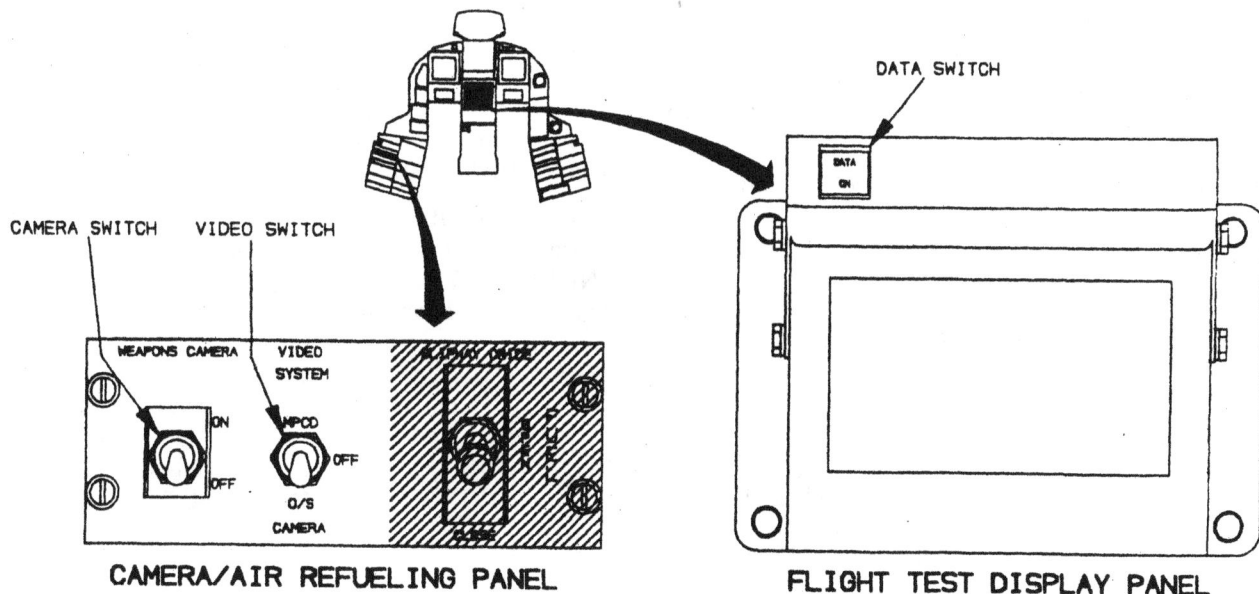

**Figure 1-69**

of day. If neither display is being recorded, the HUD is recorded and the time of day appears on the HUD.

MPCD records the information shown on the selected displays. To record a display that does not have the label RCD, a display with the record function must be selected, the record function activated, and then the display to be recorded must be reselected.

During split screen recording of both MPCD, pressing the trigger switch to the first detent provides a momentary recording of the HUD. If either the trigger switch is pressed to the second detent or the weapon release button is pressed, the HUD is recorded for ten seconds after switch release.

| | |
|---|---|
| MPCD | Data from the MPCD or HUD is recorded. |
| OFF | System turned off. |
| O/S | turns the over the shoulder camera on. |

### VMS TEST PANEL

The VMS test panel (figure 1-70) contains controls that vary the input devices and/or schedule the VMS functions.

### BLC Select Switch

The three-position, lever-locked BLC select switch allows manual positioning of the boundary layer control doors. With the switch in the unlabeled center position, doors are controlled automatically by the VMS computers. With the switch in LEFT or BOTH, the doors are controlled by the flaps/BLC switches.

| | |
|---|---|
| LEFT | Manual control of left BLC doors only. |
| BOTH | Manual control of left and right BLC doors. |

### Flutter Exciter Control Unit Switch

The three-position, lever-locked flutter exciter control unit (FECU) switch is used for flutter testing.

| | |
|---|---|
| MIXER | Flutter exciter control unit inputs are entered at the control surface mixing logic of the flight control laws software. |
| STICK | Flutter inputs from the FECU are entered where the pilot control stick/rudder pedal |

command inputs are made to the control laws.

| | |
|---|---|
| N/A | No designated function. |

### Stall Margin Switch

The three-position, lever-locked stall margin switch, labeled ISM/VSM, provides increased engine stall margin but decreased engine performance.

### BLC Accelerate Switch

The two-position, lever-locked BLC accelerate switch selects one of two predetermined actuation schedule to dynamically position boundary layer control doors according to engine airflow and Mach number.

| | |
|---|---|
| NORM | BLC doors are positioned by a predetermined schedule for normal flight. |
| ACCEL | BLC doors are positioned by a predetermined schedule to optimize performance. |

### INS Simulated Fail Switch

The two-position, lever-locked INS EFAIL switch is used to block INS Euler angle information to the control laws. The INS is not turned off.

| | |
|---|---|
| NORM | INS Euler angle computation is sent as an input to the control laws. |
| EFAIL | A value of zero is sent to the control laws in lieu of INS Euler angle information. |

### Flaps/BLC Enable Switch

The two-position, lever-locked flaps/BLC enable switch enables the switches, labeled LEF/BLC INNER, TEF IB/BLC OUTER, and TEF OB to manually position flaps or BLC doors.

| | |
|---|---|
| FLAPS | • Switch labeled LEF/BLC INNER manually positions leading edge flaps.<br>• Switch labeled TEF IB/BLC OUTER manually positions inboard trailing edge flaps.<br>• Switch labeled TEF OB/A9 manually positions outboard trailing edge flaps. |

## VMS TEST PANEL

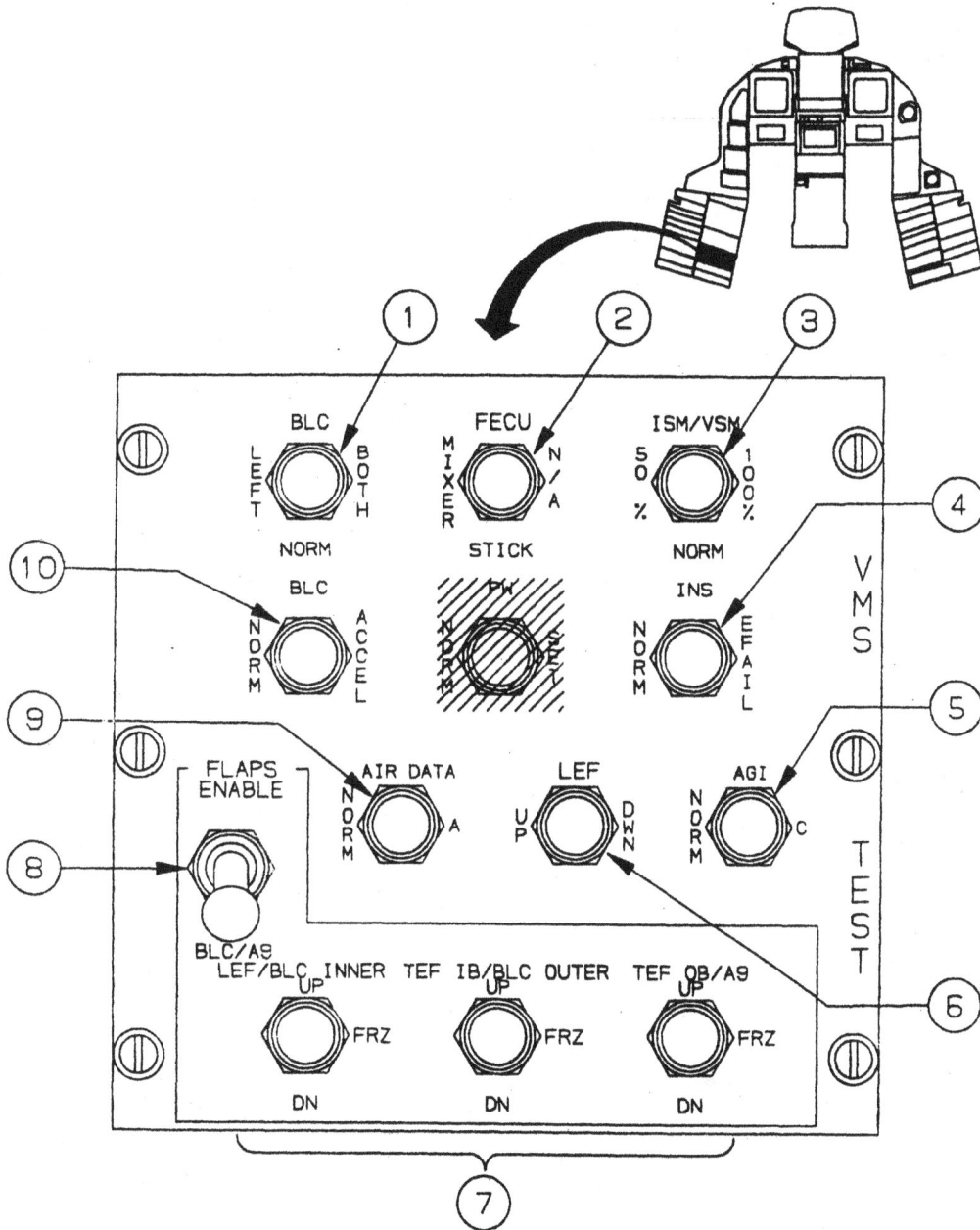

1. BLC SELECT
2. FLUTTER EXCITER CONTROL UNIT
3. INCREASED STALL MARGIN
4. INS SIMULATED FAIL
5. ARMAMENT GAS INGESTION
6. LEADING EDGE FLAPS
7. FLAP/BLC SWITCHES
8. FLAP/BLC ENABLE
9. AIR DATA
10. BLC ACCELERATE

A23-096

**Figure 1-70**

| BLC/A9 | Directs manual inputs to left or both BLC doors depending on BLC select switch position. |
|---|---|

- Switch labeled LEF/BLC INNER manually positions **inboard** BLC door(s).
- Switch labeled TEF IB/BLC OUTER manually positions outboard BLC door(s).
- Switch labeled TEF OB/A9 manually opens/closes the A9 exhaust nozzle.

### Flaps/BLC Switches

All three flaps/BLC switches, labeled LEF/BLC INNER, TEF IB/BLC OUTER, and TEF OB/A9 are spring-loaded to the center freeze position.

| UP | Either open BLC doors/A9 exhaust nozzles or retract flaps, depending on position of flaps/BLC enable switch. |
|---|---|
| FRZ | Holds BLC doors/flaps/A9 exhaust nozzle in last position commanded. |
| DN | Either closes BLC doors/A9 exhaust nozzles or extend flaps, depending on position of flaps/BLC enable switch. |

### Air Data Switch

The two-position, lever-locked air data switch directs the air data computer to use correction tables for Pitot tubes or flush mounted total pressure port. The Pitot static plumbing must correspond to the selected option.

| NORM | Uses correction schedule for L-shaped Pitot tubes. |
|---|---|
| A | Uses correction schedules for nose mounted total pressure port. |

### Leading Edge Flaps Switch

The three position leading edge flap toggle switch manually controls flap position.

| UP | Retracts leading edge flaps. |
|---|---|
| Center | The leading edge flaps are controlled by the VMC. |
| DWN | Extends leading edge flaps. |

### Armament Gas Ingestion Switch

This switch is not functional.

# FLUTTER EXCITER CONTROL UNIT

The flutter exciter control unit (FECU) generates oscillatory signals to the VMS for control system activation to determine flutter, stability, and control characteristics. The unit replaces the left MPCD for designated missions (figure 1-71). The right MPCD must be enabled to show system cautions. The VMS test panel switch labeled FECU selects either the control stick or the flight control actuators for the input of excitation signals.

The flutter exciter control unit consists of a power switch, display screen, data entry buttons, excitation modes buttons, control selector, phase switch, and operation buttons.

### POWER SWITCH

The power switch turns the flutter exciter control unit on and runs a start-up BIT. The position OFF may be used to abort a flutter exciter run, if required.

### DISPLAY SCREEN

The display screen provides ten data sets on five lines. The left side of the screen are selectable sets of data to be shown or controlled.

The right side of the screen shows the parameters for the selected program, depending on the excitation mode selected.

| START | The start frequency for the excitation. |
|---|---|
| FREQ | The present frequency of excitation. |
| STOP | The frequency used to stop excitation. |
| % AMP | The control surface displacement amplitude in percent of full control authority. Amplitude varies from 1 % to 99 % in 1% increments. |
| TIME | • Time in milliseconds for the frequency duration of a sweep program to run. Time varies from 10 to 1000 milliseconds in 10 millisecond increments.<br>• Total length of time for a random or dwell program to run. |

## FLUTTER EXCITER PANEL

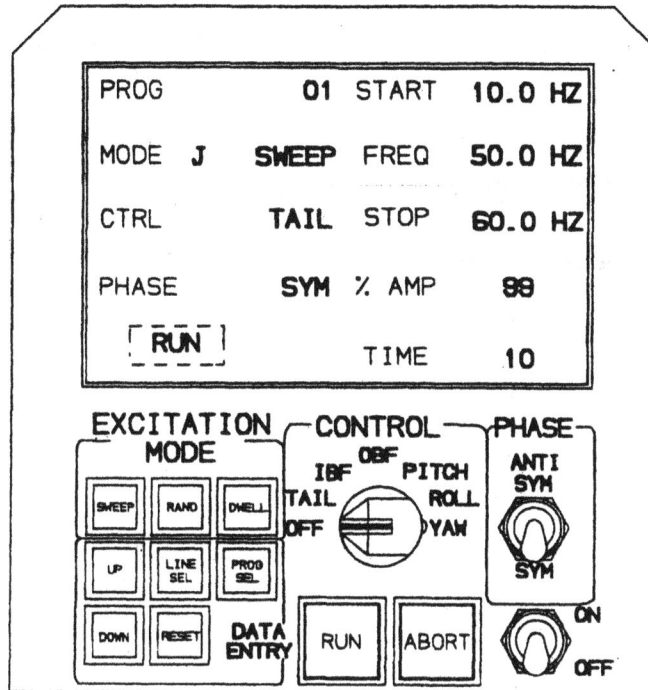

A2B-212

**Figure 1-71**

### DATA ENTRY

The data entry buttons are used to change the selected program and the program parameters shown on the display screen.

#### Program Select

The program select button labeled PROG SEL sequentially steps to programs 01 thru 15. The number of the program is shown on the display screen adjacent to the label PROG. The program parameters are listed on the right side of the display screen for each program selected.

#### Line Select

The line select button labeled LINE SEL is used to change a parameter of a program. A caret mark on the right edge of the display screen identifies the parameter to be changed and steps down to the next changeable parameter when the line select is pressed again.

After the parameter is selected with the line select button, the button labeled UP increases the parameter value or the button labeled DOWN decreases the parameter value. The button labeled RESET changes the parameter back to the original value provided the line select button has not been pushed.

### EXCITATION MODES

The control surfaces can be excited by three flutter modes. The flutter modes use program parameters to control surface oscillation.

| | |
|---|---|
| SWEEP | Sweeps control surface oscillation from the start frequency to the stop frequency. |
| RAND | Control surface oscillation is random varying from 1 Hz to the stop frequency. The start frequency is blank. |
| DWELL | Control surface oscillation dwells at the preset frequency. Start and stop frequencies are blank. |

### CONTROL

An excitation program can be accomplished on the tail, outboard flap, and inboard flap control surfaces. Positions PITCH, ROLL, and YAW are inactive.

| | |
|---|---|
| OFF | Control surface not designated. |

| TAIL | Flutter program run on tail control surfaces. |
|------|-----------------------------------------------|
| OBF | Flutter program run on outboard flap control surfaces. |
| IBF | Flutter program run on inboard flap control surfaces. |

## PHASE

The designated control surfaces can be oscillated symmetrically or asymmetrically.

| ANTI SYM | Left and right control surfaces are oscillated 180 degrees out of phase. |
|----------|--------------------------------------------------------------------------|
| SYM | Left and right control surfaces are oscillated in phase. |

## OPERATION BUTTONS

The operation button labeled RUN, starts the selected program on the designated surfaces with a symmetrical or asymmetrical phase. The button labeled ABORT can be used to abort a run. The flutter exciter control unit power switch or the stick paddle switch could also be used to abort a run. When the paddle switch is used to abort a run, the paddle must be held until the abort button is pressed or the power switch is turned to OFF.

# [1] ARMAMENT SYSTEM

The armament system consists of an internal missile carriage system and provisions for an internally mounted gun. The aircraft can carry AIM-120A (AMRAAM) and has provisions to carry the AIM-9 (Sidewinder) missile. The stores management system is part of the avionics system dedicated to weapons status and employment. The stores management system controls the weapons bay doors and weapons firing. A consent signal from the VMS, indicating adequate hydraulic pressure, is necessary for the stores management system to open/close weapon bay doors and fire weapons. The armament system described is the configuration after the weapons integration enhancement. Some hardware might not be installed and weapon switches described may not be functional prior to the enhancement.

## INTERNAL WEAPONS CARRIAGE SYSTEM

The internal weapons carriage system consists of the weapons bay insert, advanced technology launcher(s), the door launch assembly, and the weapons bay doors and drive mechanism. Each hydraulically powered door is mechanically linked to an airflow spoiler. The spoiler extends six inches below the moldline when deployed.

### Weapons Bay Insert

The weapons bay insert supports the advanced technology launcher and a modified LAU-106 for captive carriage of an AIM-120A. The platform, suspended at an outward angle of 17 degrees, imparts a down and out trajectory to missiles ejected by the launcher.

### Advanced Technology Launcher

The electrically actuated, hydraulically powered advanced technology launcher has the capability to eject launch missiles. The launcher consists of a launcher mechanism and the hydraulic power unit. The hydraulic control unit uses two accumulators that are charged by the aircraft hydraulic system after gear retraction. The programmable armament control set provides the control input to the hydraulic unit to extend or retract a linkage arm. Sensors monitoring accumulator charge, launcher position, door position, and confirmation of ejection provide the programmable armament control set with information to initiate various sequences. After the missile is launched, the launcher automatically retracts. If the missile is hung, the launcher remains in the extended position.

The advanced technology launcher ejects the AIM-AIM-120A by releasing the missile at the end of the launcher linkage arm extension stroke. The missile is aligned 4.5 degrees nose down on release with a slight nose-down momentum.

The launcher also has provisions to rail launch AIM-9 missiles with the installation of an LAU-114 launcher.

### Door Launch Assembly

Structural provisions are included to mount a weapons bay door launch assembly for AIM-9 carriage and release.

### GUN SYSTEM

Provisions are included for a M61A1 20mm gatling gun in the right side of the fuselage. Blank structural doors replace the louvered doors and blast deflector.

## ARMAMENT SYSTEM CONTROLS AND DISPLAYS

### Armament Control Panel

The armament control panel (figure 1-72) has three lever-locked toggle switches for selecting and activating armament system functions.

### MASTER ARM SWITCH

The two-position, lever-locked to SAFE master arm switch, controls electrical armament bus power for all weapons firing/launch/release signals.

| | |
|---|---|
| ARM | • Provides electrical power for weapons firing. Gun cross appears on the HUD.<br>• Powers weapon launch control circuits and enables servo controller operation to open/close weapons bay doors. |
| SAFE | • Removes power from the weapon launch control circuits and inhibits servo controller operations.<br>• Removes gun cross from HUD. |

### WEAPONS BAY DOOR SPEED SWITCH

The three-position weapons bay door speed switch selects one of three opening and closing speeds for the weapons bay doors. The switch is lever-locked in each position. Use of the high speed (NORM) and

## ARMAMENT CONTROL AND SAFETY PANELS

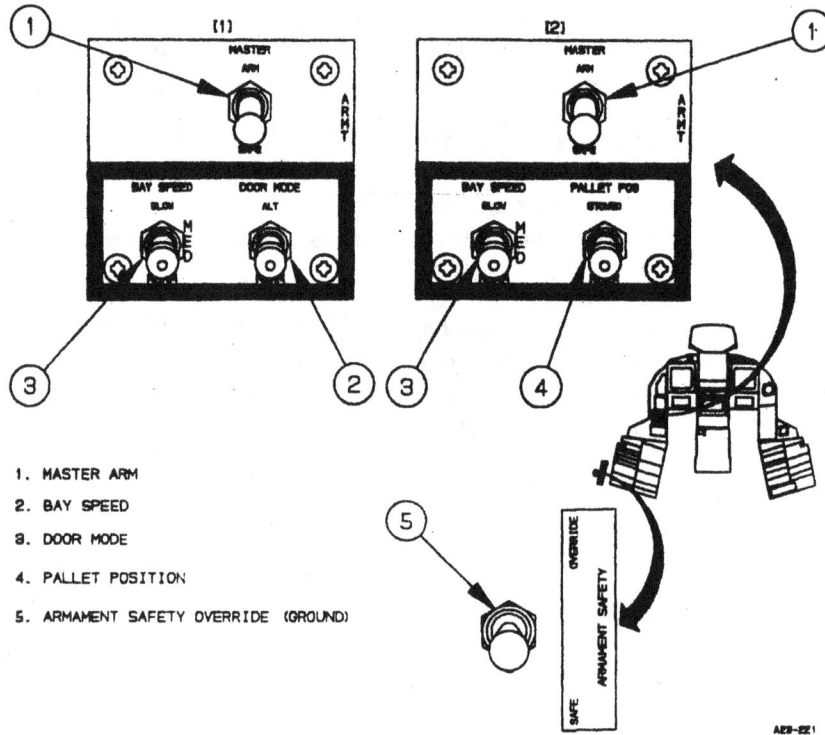

**Figure 1-72**

1. MASTER ARM
2. BAY SPEED
3. DOOR MODE
4. PALLET POSITION
5. ARMAMENT SAFETY OVERRIDE (GROUND)

medium speed (MED) modes are normally inhibited on the ground.

| | |
|---|---|
| NORM (normal) | Doors open or close in 2.1 seconds. Used for automatic launcher/door sequencing AIM-120A launch. |
| MED | Doors open or close in 5 seconds. |
| SLOW | Doors open or close in 10 seconds. Primarily for ground use. |

### DOOR SWITCH

The two-position door switch enables the throttle switch to open and close the armament bay doors without arming weapons. The switch is lever-locked in each position. This switch is inoperative until completion of weapons integration enhancement.

| | |
|---|---|
| ALT | Provides power to the bay door switch on the left throttle grip with the master arm switch in SAFE. |

| | |
|---|---|
| NORM | The bay door switch is not powered unless the master arm switch is set to ARM. |

### Armament Safety Override Switch

The two-position armament safety override switch is used to override the armament safety interlocks when the aircraft is on the ground. The switch is solenoid-held in OVERRIDE until electrical power is removed, the landing gear handle is raised, or the switch is manually placed in SAFE.

| | |
|---|---|
| SAFE | Normal ground safety interlocks for the armament system apply. |
| OVERRIDE | Bypasses the ground safety interlocks. |

### Throttle -- Armament Control

### WEAPON SELECT SWITCH

The three-position weapon select slide switch on the right throttle (see Throttles, this section) selects the type of weapon to be employed. The forward position selects the AIM-120A medium range missile (MRM), the center position selects the AIM-9 or short range

range missile (SRM), and the aft position selects the gun (if installed).

## BAY DOOR SWITCH

The three-position, spring-loaded to center position bay door switch, on the outboard side of the left throttle, (see Throttles, this section) positions the armament bay doors.

| | |
|---|---|
| Up | Closes bay doors if the launcher is in the retracted position. |
| Center | Off, or neutral position. |
| Down | Opens bay doors. |

### Control Stick -- Armament Control

#### WEAPONS RELEASE BUTTON

The control stick weapon release button launches the selected missile with the system armed and puts an event marker on the HUD for recording.

#### GUN TRIGGER

The control stick trigger has two detents. The first detent starts the HUD camera and puts an event marker on the HUD for recording. The second detent fires the gun (if installed).

## ARMAMENT DISPLAYS

The MPCD shows the weapon status and the setting of each armament system control.

### Armament Control Display -- AIM-120A

The armament control display (figure 1-73) is selected by the pushbutton labeled ARMT on the MPCD menu. The information appearing on the display varies according to the selection of weapons (position of the weapons select switch). The display with AIM-120A missile selected is described. The AIM-9 missile and gun are not covered.

Figure shows a typical display with MRM selected. The legend 120A indicates the missile loaded. The legend STBY, below the legend 120A indicates power has been applied to the missile. A boxed RDY in this field indicates the VMS has sent armament consents and missile is armed and ready to be fired. A blank indicates missile launched. A legend D below the legend STBY indicates a dummy missile. The legend HUNG in the field indicates a hung missile during the firing sequence. The legend MASTER ARM or MASTER SAFE reflect the position of the master arm switch.

The legend MRM SELECTED, SRM SELECTED, or GUN SELECTED indicates the position of the weapon select switch.

Entries under legend STATUS reflect position of weapons bay doors and launcher. The position of the doors are indicated by the legend DOORS CLOSED or DOORS OPEN. Launcher status is LNCHR DOWN, LNCHR STOWED, or LNCHR READY (Launcher accumulator pressure, volume, and temperature are satisfactory for eject launch.). LNCHR XXXX indicates a transition.

The legend SPEED NORM, SPEED MED, or SPEED LOW indicate the position of the bay speed switch. The legend DOOR MODE NORM or DOOR MODE ALT indicate the door mode switch position.

### Missile Launch -- AIM-120A

With MRM selected, alternately pressing the pushbutton adjacent to the weapon display legend TM PWR turns AIM-120A telemetry on/off. Telemetry on is indicated by a box around TM PWR.

Either a manual or an auto launch can be performed. Aiming is accomplished visually by placing the target within the HUD target designation circle.

#### MANUAL LAUNCH

The doors are opened by pressing the bay door switch to the down (open) position. When the doors are open, master arm switch ARM, and launch conditions satisfied, the legend STBY changes to RDY. The door and launcher status indicates doors open and launcher ready. The weapons release button releases the missile. A successful ejection of the missile is indicated by the legend RDY going blank and the launcher retracting. The doors are closed by raising the bay door switch to the up (close) position.

#### AUTOMATIC LAUNCH

With the system armed (legend RDY appears), the weapons release button starts the automatic launch sequence. The doors open and the launcher extends, ejecting the missile. After the missile is launched, the launcher retracts, the doors close, and the legend RDY goes blank. If the missile is hung, the launcher and doors remain extended, and the legend HUNG appears below RDY.

## ARMAMENT SYSTEM CAUTION

The armament system has one caution.

| | |
|---|---|
| ARMT (MPCD) | Weapons bay doors, spoilers, or launcher not in the commanded position. |

# ARMAMENT CONTROL DISPLAY -- AIM-120 (TYPICAL)

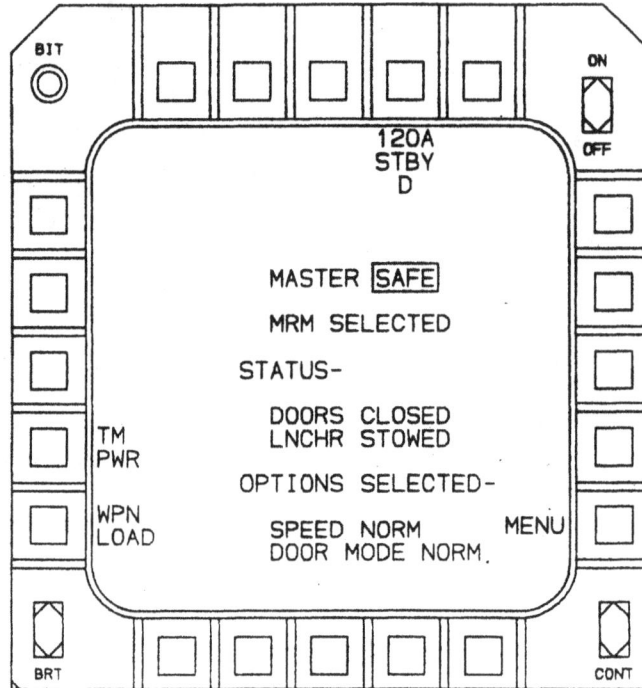

BIT     ON / OFF

```
                    120A
                    STBY
                     D


          MASTER  SAFE

          MRM SELECTED

       STATUS-

              DOORS CLOSED
  TM          LNCHR STOWED
  PWR
       OPTIONS SELECTED-

  WPN
  LOAD        SPEED NORM        MENU
              DOOR MODE NORM.
```

BRT     CONT

A28-222

*Figure 1-73*

# SECTION II

## NORMAL PROCEDURES

B23-SECII

# TABLE OF CONTENTS

## PREPARATION FOR FLIGHT

### TAKEOFF AND LANDING DATA CARD

If the takeoff distance exceeds 50 percent of the available runway, the takeoff and landing data card in the Aircrew Checklist should be completed.

### WEIGHT AND BALANCE

For weight and balance information refer to DD form 365-4 (FORM F), the FOLD, and TO 1-1B-40.

### PREFLIGHT CHECK

1. Aircraft Forms -- Check.
   Check form for aircraft status and release.

2. Takeoff and Landing Data -- Compute.

## EXTERIOR INSPECTION

- Check general condition of aircraft.
- Aircraft exterior should be checked for abnormalities that could affect flight (e.g. cracks

or leaks). Access panels could be open for after start servicing or preflight checks. Check all sensors (Pitot tubes (2), and air data ports (5), and total air temperature (2). Check doors and panels closed and fastened. Check tires for condition and inflation. Check gear struts for extension, gear pins (3) and canopy pin removed.

● Check ground safety switch safe. Ground safety switch in right main gear wheel disables the emergency power unit when installed.

## BEFORE ENTERING COCKPIT

1. Canopy -- Full open.
   External canopy switch on left side of nosewheel well must be held in OPEN position until canopy is completely open.

2. Canopy Indicator -- Not fired.
   If gray cap and orange spring are exposed, initiator has fired.

3. Canopy Jettison Safety Pin -- Removed.

4. Ejection Seat
   Check all seat actuator pins removed.

   a. Ejection Handle Safety Lever -- Lock (up).

   b. Safety Pins (2) -- Removed.

   c. Seat Kit Selector -- AUTO.

   d. Radio Beacon Selector -- AUTO.

   e. Restraint Emergency Release Handle -- Down.

   f. Battery Window -- No red showing.

   g. Seat Hose Quick Disconnect Coupling -- Check secure.

   h. Shoulder Straps -- Check.

   i. Emergency Oxygen Bottle Pressure -- Check.

   j. Pitot Tubes -- Check.

   k. Seat and Canopy Interconnect Lanyard -- Secure and safety wired.

## INTERIOR INSPECTION

A thorough cockpit interior preflight shall be accomplished before each flight. Switch positions designated AS REQUIRED indicate switch or control position may vary.

### Life Support Equipment

1. Harness And Personal Equipment Leads -- Connected.
   Attach parachute riser straps to harness fittings. Attach survival kit. Secure and adjust lap belt. Connect oxygen, jerkin/helmet cooling, anti-G suit and communication leads. Check operation of shoulder harness locking mechanism.

### Left Console

1. Armament Safety Override -- SAFE.

2. Anti-G Switch -- NORMAL.

3. Intercom Panel

   a. Mode 4 Code -- As required.

   b. Mode 4 Reply -- As required.

   c. IFF Master -- NORM.

   d. Mode 4 Crypto -- NORM.

   e. Intercom -- ON.

   f. UHF Antenna -- As required.

   g. Intercom Volume -- As required.

   h. TACAN Volume -- As required.

4. VMS Test Panel

   a. Flaps/BLC Switches (3) -- FRZ.

   b. Flaps Enable Switch -- BLC/A9.

   c. Air Data -- NORM.

   d. Leading Edge Flaps -- As required.

   e. Armament Gas Ingestion -- NORM.

   f. BLC Accelerate -- NORM.

   g. PW -- NORM.

   h. INS -- NORM.

   i. BLC Select -- NORM.

   j. FECU -- STICK.

   k. ISM/VSM -- NORM.

5. APU -- OFF.

6. EPU -- OFF.

7. Flight Test Functional Control Panel -- See data card.

8. VMS Controls

    a. Gain Enable -- OFF.

    b. Takeoff Trim Button -- Up.

    c. Fixed Gain -- OFF.

    d. VFT -- OFF.

    e. VMS BIT Consent -- OFF.

    f. Flap Emergency -- NORM.

    g. Flight Control Override -- OFF, guard closed.

    h. Yaw Trim -- OFF.

    i. VMS/ENG Reset Switch -- OFF.

9. Weapons Camera -- Off.

10. Video Control System Switch -- Off.

11. Air Refueling Switch -- CLOSE.

12. Throttles -- OFF.

13. Weapons Bay Door Switch -- Center.

14. Speedbrake Switch -- Forward.

15. Engine Mode Lever -- CONV.

16. Throttle Friction -- Set.

17. Lights -- Set.

    a. Interior Lights Test Switch -- Off.

    b. Anticollision Light -- ON.

18. Ground Power Panel Switches (4) -- AUTO.

19. Brake Control Switch -- MECH.

20. Antiskid -- NORM.

21. Canopy Jettison Lever -- Forward.

**Instrument Panel**

1. Landing Gear Emergency Extension -- Norm, guard closed.

2. Brakes/NWS Emergency Control -- NORM.

3. Bay Speed Switch -- SLOW.

4. [1] Door Mode -- NORM.

5. Master Arm -- SAFE.

6. Landing Gear Handle -- DN.

7. Fire Pushbuttons (3) -- Out, not depressed.

8. Fire Extinguisher Discharge Switch -- OFF.

9. UPFRONT Control

    a. Radio Volume Controls (2) -- As required.

10. HUD

    a. Brightness Knob -- Off.

    b. Declutter Switch -- As desired.

    c. HUD Intensity -- DAY.

11. Standby Flight Display Selector -- 1.

**Right Console**

1. Electrical Panel

    a. Generator (L and R) -- ON.

    b. Emergency Generator Switch -- ON.

    c. VMS Batteries (2) -- OFF.

    d. Emergency Battery -- OFF.

    e. Utility Battery -- OFF.

2. Oxygen Regulator

    a. Jerkin Switch -- OFF.

    b. Oxygen Supply -- OFF.

    c. Oxygen Switch -- NORM.

    d. Oxygen Power -- OFF.

3. ECS Panel

    a. ECS Mode -- AUTO

    b. Temperature Control -- As required.

    c. Cabin Pressure -- RAM/DUMP.

    d. Anti-Ice/Pitot Heat -- AUTO.

    e. Bleed Air Control -- BOTH.

4. Defog/Cabin Air Lever -- CABIN AIR.

5. INS Selector -- OFF.

6. Jerkin Thermal Control -- OFF.

7. Utility Light -- Stowed.
Set power mode and fill selectors as required. Turn volume control full clockwise.

8. KY-58 Panel -- Set.

# POWER ON

1. External Cooling Air -- Connected.

2. External Electrical Power -- Connected.

3. Utility Battery -- EXT PWR (2 seconds), then ON.

4. Emergency Battery -- ON.

5. VMS Batteries (2) -- ON.

6. Ground Power Panel Switches

   a. MPDP -- ON.

   b. Switch 1 -- B ON.

   c. CC -- ON.

   d. Switch 2 -- ON.

7. MPCD (2) -- On.
   Verify caution BATTERY does not appear.

8. HUD -- On.
   Set brightness knob to mid position.

9. Radios (2) -- On.

10. IFF -- ON, modes as required.
    IFF must be on with one mode (minimum) selected to complete AUTO BIT checks.

11. TACAN -- ON, set as required.

12. Built-In Test
    Approximately eight seconds after avionics are powered up, false failures appear on the BIT display:

    • MPDP A6A
    • MPDP B6A
    • MPDP A6B
    • MPDP B6B
    • L FEED
    • R FEED
      The L FEED/R FEED false failures clear during the fuel low portion of the avionics automatic BIT.

    a. MPCD MENU, BIT, AUTO Buttons -- Press.

    b. Wait -- 2 minutes.
       After AUTO BIT is complete (approximately 2 minutes), note results.

    c. MAINT, AUDIT, CC, RESET Buttons -- Press.
       To reset the MPDP, select AUDIT and step to the CC option from the maintenance display and press the button labeled RESET.

13. Intercomm -- Check.
    Establish communications with ground crew.

14. Standby Flight Display -- Test, then 1.
    Rotate Selector to test position. Check test format for missing segments or indicators.

15. INS
    Accomplish a gyrocompass or a stored alignment. Do not move the aircraft during INS alignment. After approximately 60 second delay, NO TAXI disappears from HUD. GC quality numeric appears next to GC on HUD. Quality number decreases to between 1.2 and 0.7. Alignment is complete when GC OK is displayed on HUD (approximately four minutes).

    a. Selector -- ALIGN-GC/STORE.

    b. UFC Present Position -- Check.
       Select present position and update submenu on UFC. Enter or verify present position coordinates. Select INS Data Source. Asterisk appears next to INS.

16. HUD Titling

    a. UFC PP And TITL Buttons -- Press.
       Select HUD titling submenu from present position menu on UFC.

    b. Data -- Enter (see data card).
       Enter month, day, year, mission number, and aircraft number.

    c. Video System Control Switch -- MPCD (hold 10 seconds), then OFF.

17. UFC Flight Data
    Enter test card data on UFC menus.

    a. Bingo Fuel -- As required.

    b. QNH -- Enter.

    c. Envelope Data -- Enter.
       See flight test card. Enter calibrated air speed, Mach number, angle of attack, angle of sideslip, aircraft weight, CG, G, and aft CG data.

    d. Weight and CG Data -- Load and verify.
       To load weight and CG data into the vehicle management computers, simultaneously press control stick trigger to second detent and select VMS BIT consent. Verify the calculated CG on the standby flight display.

    e. Navigation Steer Point Data -- Enter.

18. Interior Lights Test Switch -- Test.
    Ensure all warning and caution lights come on.

## BEFORE STARTING ENGINES

WARNING

With engines running, and without proper ear protection, physical injury and hearing loss can occur. Wear maximum ear protection possible. Wear ear plugs in addition to pilot helmet.

1. Danger Areas Fore and Aft -- Clear.
   Air intake danger area is from 5 feet aft of the intake, out 25 feet to the side, and a 25 foot arc in front. For aft danger areas, see engines section I or systems safety supplement to Section V.

2. Fire Guard -- Posted.

3. INS Selector -- As required.
   After successful alignment (GC OK displayed on HUD), select NAV. If stored alignment is planned, select OFF.

4. Fire Warning -- Test.
   Check the left and right engine fire, and APU fire warning lights come on. Listen for voice warnings.

5. Bleed Air Control -- Cycle OFF, then BOTH.
   L (R) BLEED advisories go out.

6. MPCD Engine Display -- Select and record.
   Monitor engine parameters on right MPCD.

   a. Video System Control Switch -- MPCD.

   b. RCD Button -- Press.

7. Flight Test Data Mode -- DATA ON.

8. Throttles -- MAX, then OFF.
   Slowly advance throttles individually from OFF to IDLE to MAX to OFF. Verify telemetry.

## EXTERNAL AIR START

Engines can be started using an external air source, crossbleed air from an operating engine, or the APU (if installed). Electrical power can be supplied by an external electrical source or the utility battery. Monitor engine parameters on standby flight display.

The right engine is normally started first to check for proper utility hydraulic pressure readings. The right utility hydraulic pressure is approximately 100 PSI lower than the left side.

To avoid the potential of foreign object damage, external electrical and external cooling air are normally disconnected after the right engine is started.

For engine start parameters, abort engine start criteria, and sequence of engine start, see Engines, Section I.

If engine does not light off within limits, abort start, and motor the engine (if required).

1. External Starting Air -- On.
   Coordinate with crew chief.

2. Engine Crank -- R.

3. Right Throttle -- IDLE.
   At appropriate RPM or time, lift finger lift and push throttle forward to the idle position.

4. Engine Indications -- Check.
   Check hydraulic pressure.

5. VMS/ENG -- RESET.

6. Caution R GEN -- Check out.

7. Oxygen Monitor Test Button -- Press and hold.
   Check warning light OXY and master caution light come on within 15 seconds. Light should go out within 1 minute after releasing button.

8. Oxygen Regulator

   a. Jerkin Switch -- As required.
      If a jerkin is not worn, the oxygen regulator jerkin switch shall be in OFF. If a jerkin is worn, the jerkin switch must be ON to test the jerkin system and to provide G-biased pressures.

   b. Oxygen Supply -- ON.

   c. Oxygen Power -- 100%, then NORM.
      Check increased pressure in oxygen mask while switch is in 100%.

   d. Flow Indicator -- Check.
      Take a few deep breaths and check flow indicator flashes on and off to indicate an oxygen flow. Hold breath and indicator light remains out.

   e. Oxygen Switch -- TEST (hold).
      Check mask seal against the increasing pressure. If the jerkin is worn, the jerkin, helmet bladder, and the anti-G suit inflate.

   f. Oxygen Switch -- EMERG, then NORM.
      Check increased pressure while the switch is in EMERG.

9. External Cooling Air and Electrical -- Disconnect.
Cockpit cooling airflow stops until canopy is closed.

10. Cabin Pressure Switch -- NORM.

**See CROSSBLEED START for left engine start, or:**

11. Engine Crank -- L.

12. Left Throttle -- IDLE.
At appropriate RPM or time, lift finger lift and push throttle forward to the idle position.

13. Engine Indications -- Check.
Check hydraulic pressures.

14. VMS/ENG -- RESET.

15. External Starting Air -- Off.

## CROSSBLEED START

> **WARNING**

For crossbleed start, the operating engine is required to be above idle RPM. Ensure all personnel are clear, both fore and aft of the aircraft.

1. Bleed Air Control -- BOTH.

2. Operating Engine -- RPM as required.

3. Engine Crank -- Opposite side from operating engine.

4. Throttle -- IDLE.
At appropriate RPM or time, lift finger lift and push throttle forward to the idle position. Use the appropriate engine specifications.

5. Engine Indications -- Check.
Check hydraulic pressures.

6. Engine Crank -- Check OFF.

7. VMS/ENG -- RESET.

8. Both Throttles -- IDLE.

## BEFORE TAXI

1. Both Generators -- Check On.
Check cautions L (R) GEN out.

2. Right Generator Switch -- OFF, RESET, then ON.
Cycle the right generator to off, then reset to verify the left generator accepts the electrical

load. Both MPCD will go dark if electrical load is not accepted.

3. Jerkin Thermal Control -- As required.
If used, the temperature selector should be set out of OFF to a comfortable temperature. Check lighted legends NO FLOW and INOP are off.

4. Flight Test Data Button -- DATA OFF.

5. MPCD Recorder

   a. Video Control System Switch -- OFF.

   b. RCD Button -- Press.

6. Canopy -- As required.

7. Anti-G Suit Valve -- TEST.
Anti-G suit inflates. If a jerkin is worn and the oxygen regulator jerkin switch is in ON, the jerkin inflates.

8. VMS BIT -- Complete.
Reset VMS/ENG to clear cautions, then simultaneously hold VMS BIT consent switch in CONSENT and press MPCD VMS BIT button and release. VMS BIT runs for approximately 20 seconds. Check VMS BIT display for test results. Confirm with the flight test director that the air data system completed and passed IBIT. VMS/ENG reset, as required.

9. Speed Brake Switch -- Forward.

10. Flight Controls and Trim

    a. Flight Test Data Button -- DATA ON.
    Wait 15 seconds after legend DATA ON comes on.

    b. Takeoff Trim Button -- Press.
    Press button and hold until light in button comes on.

    c. Trim -- Cycle in each axis.

    d. Takeoff Trim Button -- Press.
    Press button and hold until light in button comes on.

    e. Flight Controls -- Cycle.
    Check with ground crew that flight controls are clear. Move flight control stick slowly through all three axis.

    f. Flight Test Data Button -- DATA OFF.

11. MPCD (L and R) -- Program.
Check fuel and armament displays. Select desired displays for takeoff.

12. Master Arm -- ARM.

13. [1] Bay Speed Switch -- SLOW.

14. [1] Weapons Bay Doors -- Close.
Confirm with ground crew that the area is clear before closing weapons bay doors. Crew chief must simultaneously hold ground weapons bay door switch to CLOSE.

15. Master Arm -- SAFE.

16. Flaps -- DOWN.

17. INS Switch -- NAV.

18. Wheel Chocks -- Removed.

## TAXI

As the aircraft begins to move, apply brakes to check operation. Check nosewheel steering in both directions. During heavy gross weight operations, make all turns at the minimum practical speed.

1. Brakes -- Test.

2. Nosewheel Steering -- Check.
Press and hold nosewheel steering button to check high gain, release to check low gain operation. Avoid abrupt nosewheel steering inputs in high gain.

3. Flight Instrument Indications -- Check.
Observe electronic HSI and ADI indications for smooth movement. Check standby attitude indicator.

## BEFORE TAKEOFF

1. Takeoff Trim Button -- Press.
Hold takeoff trim button pressed until light comes on. Light goes out 3 seconds after button is released.

2. EPU Switch (ground service panel) -- ARM.
Confirm with ground crew that the EPU switch in the right gear well is in ARM to enable EPU operation during flight.

3. EPU Switch -- AUTO.
After ground crew is clear of right gear well, place switch to AUTO to enable EPU during flight.

4. Engine Mode Lever -- CONV.

5. Ejection Safety Lever -- ARMED (down).

6. Warning And Caution Lights -- Check out.
Check warning and master caution lights are out. Check MPCD clear of cautions.

7. Flight Test Equipment -- See data card.
See flight test card.

8. IFF -- As required.
Set codes assigned. Select appropriate modes.

## TAKEOFF

Individually advance throttles and check instrument indications. Some aircraft creep could occur. When ready for takeoff, release brakes and advance throttles to MIL or MAX, as required. Monitor engine instruments for proper operation.

Use of high gain nosewheel steering should be limited to less than 15 knots.

During takeoff, control stick should remain in neutral position until desired rotation and then moved aft to the normal stop. Stop rotation at approximately 10 degrees pitch attitude and allow aircraft to fly off the runway.

1. Engines (L And R) -- Check.
Advance each throttle and check indications. Advancing either throttle beyond 2/3 travel cycles the flaps to the takeoff position.

2. Brakes -- Release.

3. Throttles -- As required.

## AFTER TAKEOFF/CLIMB

Retract landing gear when a positive rate of climb is verified by outside visual cues, the vertical velocity indicator, and increasing altitude.

1. Landing Gear -- UP.

2. Flaps -- UP.

3. Throttles -- As required.

**Above 8,000 feet MSL.**

4. Cabin Pressure -- Check.

## LEVEL OFF/CRUISE

Monitor aircraft systems operation throughout the flight. Frequently check engine indications, cabin pressure, and fuel quantity.

1. QNH -- Reset as required.

2. Oxygen -- Check.

# AIR REFUELING

1. Master Arm Switch -- SAFE.

2. Fuel Display -- Select.

3. IFF -- Off.
   Turn IFF off when three NM from the tanker.

4. TACAN -- REC.

5. Air Refueling Switch -- OPEN.

6. Air Refueling Ready Advisory -- AAR RDY.
   Air refueling ready advisory comes on when air refueling receptacle is open and system is ready.

## CONTACT

1. Air Refueling Ready Advisory -- Out.
   The advisory goes out five seconds after boom is engaged.

2. Fuel Quantity -- Monitor.
   Check selected tanks on the fuel display are filling.

## DISCONNECT

1. Air Refueling Disconnect Switch -- Press.

2. Air Refueling Ready Advisory -- AAR RDY.
   The advisory comes on when boom disconnects.

## POST AIR REFUELING

1. Air Refueling Switch -- CLOSE.

2. Fuel Display

   a. CG -- Check.

   b. Fuel Quantity -- Check.

3. IFF and TACAN -- As Required.

# DESCENT

Before descent, preheat the canopy by moving the defog lever forward into the defog range.

1. Weapons Bay Doors -- Check closed.

2. Master Arm Switch -- SAFE.

3. QNH -- Reset.
   Reset to local altimeter setting.

4. Fuel -- Check.
   Check fuel quantity and distribution.

5. Defog Lever -- Forward, as required.

# BEFORE LANDING

1. Approach And Braking Speeds -- Compute.

2. Landing Gear -- DN.

3. Flaps -- DOWN.

4. Speed Brake Switch -- Forward.

# LANDING

## NORMAL LANDING

Fly final approach at 10 degrees angle of attack for all aircraft configurations/conditions and 2.5 - 3.5 degrees glide slope. Flare touch down at 10 degrees AOA (8-10 knots below final approach speed) and minimum sink rate. Tail strike is possible if touchdown attitude exceeds 12.5 degrees at touchdown.

## CROSS WIND LANDING

Fly final approach with either crab or wing low into the wind but plan to touch down without crab.

## LANDING ROLLOUT

Lower nose smoothly to runway without hesitation. After nosewheel touch-down, smoothly apply full aft stick (normal detent) for maximum aerodynamic braking, maintain nosewheel on the runway.

# AFTER LANDING

1. Ejection Safety Lever -- Lock (up).

2. EPU -- OFF.
   Turn EPU selector to OFF before ground crews approach aircraft.

3. IFF -- As required.

4. Cockpit Pressure -- Check.
   Prior to opening canopy, check the cockpit is depressurized. If the cockpit is overpressurized (cockpit pressure below field elevation), place cabin pressure switch to DUMP. Attempting to open the canopy while the cockpit is pressurized could cause damage or loss of the canopy.

5. Canopy -- As required.

# SINGLE ENGINE TAXI

Shutting one engine down during taxi operations causes the left and right number 2 AC electrical buses to drop off line. Electrical power to systems powered by the two buses is lost. Adjust seat height prior to engine shutdown. To shut one engine down, first place the generator switch to OFF, then retard applicable throttle to OFF.

# ENGINE SHUTDOWN

1. INS Selector -- OFF.

2. HUD -- OFF.
   Rotate brightness knob counterclockwise to OFF detent.

3. [1] Bay Speed Switch -- Slow.

4. [1] Weapons Bay Doors -- Open.
   Confirm with ground crew that the area is clear before opening weapons bay doors.

5. External Lights -- OFF.

6. EPU -- SAFE (ground service panel).
   Confirm with ground crew that the EPU switch in the right gear well is in SAFE.

7. Mode 4 Crypto -- As required.

8. After Landing Avionics BITs -- Complete.
   Complete all avionics systems tests before power down.

9. Diagnostic Data Dump

   a. VMS/ENG Reset -- RESET and hold.

   b. VMS BIT Consent Switch -- CONSENT.

10. Fuel System Failure Audit
    Fuel system failure data is lost after engine shutdown.

    a. BIT Button -- Press.

    b. MAINT Button -- Press.

    c. AUDIT Button -- Press.

    d. FMC-M -- Select.

    e. Fuel Failure BIT (5) -- Select, enter.
       Fuel failure BIT location labels will be provided by control room.

11. Video Control System Switch -- OFF.

12. External Electrical Power -- Connected.

13. Cabin Pressure -- RAM/DUMP.
    Open pathway for external cooling air to the avionics equipment bay.

14. Canopy -- OPEN.

15. Throttle Left Engine -- OFF.

16. External Cooling Air -- Connected.

17. Throttle Right Engine -- OFF.

18. VMS Batteries (2) -- OFF.
    VMS batteries should be turned off prior to the untility battery to prevent discharge.

19. Emergency Battery -- OFF.

20. Utility Battery -- OFF.

21. External Electrical Power -- Off.

# SECTION III
## EMERGENCY PROCEDURES

C23-SECIII

# TABLE OF CONTENTS

## GENERAL

## WARNING, CAUTION, AND ADVISORY ANALYSIS

## GROUND EMERGENCIES

## TAKEOFF EMERGENCIES

# TABLE OF CONTENTS

## INFLIGHT EMERGENCIES

# TABLE OF CONTENTS

## INFLIGHT EMERGENCIES (Con't)

## AIR REFUELING EMERGENCIES

## LANDING EMERGENCIES

# GENERAL

## INTRODUCTION

Emergency procedures are grouped under ground, takeoff, inflight, and landing emergencies. Do not omit normal procedure checklists after accomplishing emergency procedures.

## WARNING, CAUTION, AND ADVISORY SIGNALS

This section contains an analysis of the warning, caution, and advisory indications. The analysis lists the individual warning, caution, or advisory message, the condition causing the message, and corrective action.

## MULTIPLE EMERGENCIES

Multiple emergencies, adverse weather, and other peculiar conditions could require modification of these procedures.

## BASIC RULES

In all emergencies, the overriding considerations must be to:

1. Maintain aircraft control.

2. Analyze the situation and take proper action.

3. Land as the situation dictates.

## DEFINITION OF TERMS

The terms land as soon as possible and land as soon as practical are defined and shall be used as general guidelines. Emergency conditions, combined with analysis of aircraft condition and type of emergency, are primary factors in determining the urgency to land.

### Land As Soon As Possible

An emergency shall be declared. A landing should be accomplished at the nearest suitable airfield considering the severity of the emergency, weather conditions, field facilities, ambient lighting, aircraft gross weight, and command guidance.

### Land As Soon As Practical

Emergency conditions are less urgent, and although the mission is to be terminated, the degree of the emergency is such that an immediate landing at the nearest adequate airfield is not necessary.

# WARNING, CAUTION, AND ADVISORY ANALYSIS

Red warning light indicators (figure 3-1), *VOICE* warning, or warning **Tone** indicate a hazardous condition requiring immediate corrective action. A yellow caution message indicates an impending dangerous condition requiring corrective action. All cautions light the master caution light and appear on the multipurpose color display. A bell tone sounds when the master caution light is displayed. A green advisory message on the MPCD is informative indication of an existing aircraft system condition or state.

The warning, caution, advisory analysis lists indications alphabetically in the first column. If the indication is presented other than on the MPCD, the means used to alert the pilot is listed below the indication in the form of an abbreviation. The condition or cause of the message is in the second column. Corrective action is listed in the third column.

Abbreviations are used in the analysis to identify the location of each warning and caution indication:
- Fire warning and controls (FW),
- Head up display (HUD),
- Warning light panel (WLP),
- Multipurpose color display (MPCD),
- *VOICE* warnings are in italic bold print.
- Warning **Tone** is in bold print.

# WCA ANALYSIS

| INDICATION | CONDITION | CORRECTIVE ACTION |
|---|---|---|
| AAR RDY (Advisory) HUD/MPCD | Air refueling system ready. | |
| AIR DATA | Air data system fault. | Reset VMS/ENG |
| L AMAD HOT R AMAD HOT | AMAD oil temperature high. | • Check fuel feed tank temperature (fuel status display). • Retard applicable engine to idle, if tank fuel temperature normal. **If feed tank fuel temperature above 60 degrees C** • Set 5,000 lbs/hr/engine fuel flow (minimum). • Land as soon as practical. |
| ANTISKID | Antiskid inoperative, off, or emergency brake selected. | See BRAKE SYSTEM MALFUNCTIONS. |
| APU ACCUM | Emergency accumulator pressure below 3400 psi. | Land as soon as practical. |
| APU/EPU FIRE *THREAT WARNING* FW | • Emergency power unit fire. • Automatic fire suppression started, if weight on wheels. | See EPU FIRE IN FLIGHT. |
| ARMT | Armament system not in commanded position. | • Slow below 250 KCAS, if required. • Recycle armament controls. |
| AUTOPILOT | Autopilot not installed. | Reset master caution light. |

# WCA ANALYSIS (Cont.)

| INDICATION | CONDITION | CORRECTIVE ACTION |
|---|---|---|
| AV BIT (Advisory) | Avionics BIT failure. | Select BIT display for cause. |
| BATTERY | Emergency, utility or one of two VMS batteries malfunction. | See BATTERY FAILURE. |
| BBW OFF | Brake-by-wire selected, but not engaged or antiskid system off. | • Check antiskid switch to NORM <br> • Reset VMS/ENG <br> **If braking not regained** <br> See BRAKE MALFUNCTIONS. |
| BINGO (Advisory) <br><br> *BINGO FUEL* | Fuel quantity at preset level. | |
| L BLEED <br> R BLEED <br> (Advisory) | Bleed air valve closed. | |
| L BLD LEAK <br> R BLD LEAK | Leak in engine bleed air system. | • See BLEED AIR LEAK -- SINGLE. <br> • See BLEED AIR LEAK -- DOUBLE. |
| L BST PUMP <br> R BST PUMP | Fuel boost pump pressure low. | • Throttles out of AB. <br> • See FUEL BOOST PUMP FAILURE. |
| BYPASS | Inlet bypass system failure. | • Maintain subsonic speed. <br> • Check BLC doors position. |
| CABIN ALT | Cabin altitude above 25,000 feet MSL. | • Check cabin pressure switch position. <br> • Check oxygen system. <br> • Descend, if required. |
| CANOPY | • Canopy unlocked. <br> • Canopy actuated initiator lanyard disconnected and normal ejection impossible. | • Slow to below 250 KCAS. <br> • Descend to below 10,000 feet MSL. <br> • Avoid excessive G loading. |
| CAUTION <br><br> Both MPCD | • Avionics interface unit 1 failure. <br> • MPCD caution messages are not shown. <br> • Master caution light is inactive. | Land as soon as practical. |
| CG | Control laws using default values. | Restrict maneuvering. |
| E GEN MAL | Emergency generator system malfunction. | Reset emergency generator. |
| ECS AIR HOT | Insufficient avionics cooling, including vehicle management system. | See ECS OVERTEMPERATURE. |

# WCA ANALYSIS (Cont.)

| INDICATION | CONDITION | CORRECTIVE ACTION |
|---|---|---|
| L ENGINE<br>R ENGINE | Engine malfunction or failure. | • Reduce power to idle.<br>• Reset VMS/ENG<br>• Check standby flight display.<br>• See ENGINE MALFUNCTIONS.<br><br>**If oil pressure is zero**<br>• Shut down engine (conditions permitting). |
| L ENG 1ST<br>R ENG 1ST | Engine fault detected. | • Check standby flight display.<br>• See ENGINE MALFUNCTIONS. |
| L ENG 2ND<br>R ENG 2ND | Engine fault detected. | • Check standby flight display.<br>• See ENGINE MALFUNCTIONS. |
| ENVELOPE<br><br>**TONE** | Exceeding selected:<br>• Airspeed.<br>• Mach.<br>• Angle of sideslip.<br>• Angle of attack.<br>• G | Correct to within envelope. |
| EPU GEN<br>(Advisory) | EPU 1 is operating. | |
| EPU HYD<br>(Advisory) | EPU 2 is operating. | |
| EPU NOT ARM | Ground EPU safety switch in SAFE.<br><br>**OR**<br><br>EPU mode selector in OFF position. | • Have EPU safety switch (ground servicing panel) put in ARM position.<br>• Check EPU mode selector in AUTO.<br>• Do not takeoff until corrected. |
| FCS 1ST<br>(takeoff roll) | Air data validity check out of tolerance. | Consider aborting takeoff. |
| FCS 1ST<br>(in flight) | Flight control system fault. | • Reset VMS/ENG.<br>• See FLIGHT CONTROL MALFUNCTIONS |
| FCS 2ND | Two flight control system like faults. | • Reset VMS/ENG.<br>• See FLIGHT CONTROL MALFUNCTIONS. |
| FCS 3RD | Three flight control system like faults. | • Reset VMS/ENG.<br>• See FLIGHT CONTROL MALFUNCTIONS. |

# WCA ANALYSIS (Cont.)

| INDICATION | CONDITION | CORRECTIVE ACTION |
|---|---|---|
| **FIRE**<br><br>**WARNING ENGINE FIRE LEFT**<br>or<br>**WARNING ENGINE FIRE RIGHT**<br><br>FW | Engine or AMAD bay fire or overtemperature. | • See ENGINE/AMAD BAY FIRE DURING START.<br>• See ENGINE/AMAD BAY FIRE IN FLIGHT. |
| FLAPS | Flap system malfunction. | • Reset VMS/ENG.<br>• See FLAP MALFUNCTIONS. |
| FUEL ASYM | Asymmetrical wing fuel distribution (500 pounds). | • Throttles out of AB.<br>• Discontinue aggressive maneuvering.<br>• See FUEL ASYMMETRY. |
| FUEL FAIL | • Fuel management computer or probe fault.<br>• Fuel quantity readings unreliable. | See FUEL MANAGEMENT COMPUTER FAILURE. |
| FUEL HOT | Engine inlet fuel temperature high. | See FUEL HOT. |
| FUEL LOW<br><br>**WARNING FUEL LOW** | Low fuel quantity in feed tank:<br>    Less than 1100 lbs in tank 1.<br>        **and/or**<br>    Less than 900 lbs in tank 2. | • Discontinue afterburner.<br>• See FUEL LOW. |
| FX GAINS | Flight controls operating in frozen gains. | See AIR DATA LOSS -- TOTAL. |
| L GEN or<br>R GEN | • Applicable generator failure.<br>• Generator control unit (GCU) malfunction. | See SINGLE GENERATOR FAILURE. |
| L GEN and<br>R GEN | Both generators failed. | • Throttles idle.<br>• See DOUBLE GENERATOR FAILURE. |
| HOLD BRAKE | Brake hold selected but not engaged. | Reset brake hold, if required. |
| HYDRAZINE | EPU is using hydrazine. | Land before hydrazine is depleted. |
| INS | • Inertial navigation system failure.<br>• Flight controls operating in degraded mode. | • Use standby attitude indicator.<br>• Avoid aggressive maneuvering. |
| NWS | • Nosewheel steering inoperative.<br>• Nosewheel steering fault detected. | • Press paddle switch and hold.<br>• Use differential braking. |

# WCA ANALYSIS (Cont.)

| INDICATION | CONDITION | CORRECTIVE ACTION |
|---|---|---|
| LH OVHT<br>RH OVHT<br><br>***WARNING OVERTEMP LEFT/RIGHT***<br><br>WLP | • Overtemperature of the deck area aft of the engine.<br>• Overheat detection system could take 60 seconds to reset. | • Terminate afterburner on affected engine.<br>• Throttle idle, if practical.<br>• Check for other indications of fire.<br>• Land as soon as practical. |
| OXY<br><br>***THREAT WARNING***<br><br><br><br><br><br>WLP | Oxygen system failure. | **Above 10,000 feet cabin altitude**<br>• Activate emergency oxygen bottle.<br>• Turn oxygen power switch off.<br>• Turn oxygen supply switch off.<br>• Descend and maintain below 10,000 feet MSL.<br><br>**Below 10,000 feet MSL**<br>• Maintain below 10,000 feet.<br>• Land as soon as practical. |
| PC-1 | PC-1 Hydraulic pressure below 1,700 psi. | • Check hydraulic pressure on standby flight display.<br>• Land as soon as practical. |
| PC-2 | PC-2 hydraulic pressure below 1,700 psi. | • Check hydraulic pressure on standby flight display.<br>• Land as soon as practical. |
| PC-1 and PC-2 | PC-1 and PC-2 hydraulic pressure low. | • See PC HYDRAULIC FAILURE -- TOTAL.<br>• See PC AND UTLITIY HYDRAULIC FAILURE. |
| RAM AIR | Emergency ram air scoop open. | • ECS cabin mode switch to OFF/RAM then AUTO.<br>• Between 20,000 to 25,000 feet MSL and 0.8 to 0.9 Mach, ram air sufficient to keep cockpit pressure at approximately 10,000 feet MSL. |
| RESET VMS | Vehicle management system fault. | Reset VMS/ENG. |
| TAIL ACT | Tail actuation system malfunction. | • Maintain below 300 KCAS.<br>• Reset VMS/ENG.<br>• Land as soon as practical.<br>• See CONTROLLABILITY CHECK |
| TANK PRESS | Fuel tank pressure low. | • Throttles out of AB.<br>• See FUEL TANK PRESSURIZATION SYSTEM FAILURE. |
| THROTTL<br>(Advisory) | • Engine not operating in mode selected by the throttle.<br>• Throttle sensor mode malfunction. | |

# WCA ANALYSIS (Cont.)

| INDICATION | CONDITION | CORRECTIVE ACTION |
|---|---|---|
| UTIL 1<br>UTIL 2 | Utility hydraulic pump output pressure below 1700 psi. | Land as soon as practical. |
| UTIL 1 and UTIL 2 | Both utility hydraulic pumps output pressure low. | • See LANDING GEAR EMERGENCY EXTENSION.<br>• See LANDING WITH UTILITY HYDRAULIC FAILURE. |
| UTIL A | • Utility A hydraulic circuit pressure below 1,700.<br>• Emergency accumulator does not recharge. | • Monitor hydraulic pressure on standby flight instrument display.<br>• Land as soon as practical. |
| UTIL B | • Utility B hydraulic circuit pressure below 1,700.<br>• Normal gear lowering, brakes, and nosewheel steering inoperative. | • See LANDING GEAR EMERGENCY EXTENSION.<br>• See LANDING WITH UTILITY HYDRAULIC FAILURE. |
| VMS BIT<br>(Advisory)<br>(ground only) | VMS BIT fault detected. | |
| VMS HOT | • Vehicle management system overheated.<br>**And/or**<br>• Engine electronic control unit overheated. | **If ENG 1ST is on**<br>• Land as soon as practical (within 30 minutes).<br>**If ECS AIR HOT is on**<br>See ECS OVERTEMPERATURE. |
| XFMR RECT | One or more of the 6 transformer rectifiers inoperative. | See TRANSFORMER RECTIFIER FAILURE. |

# WARNING AND MASTER CAUTION LIGHTS

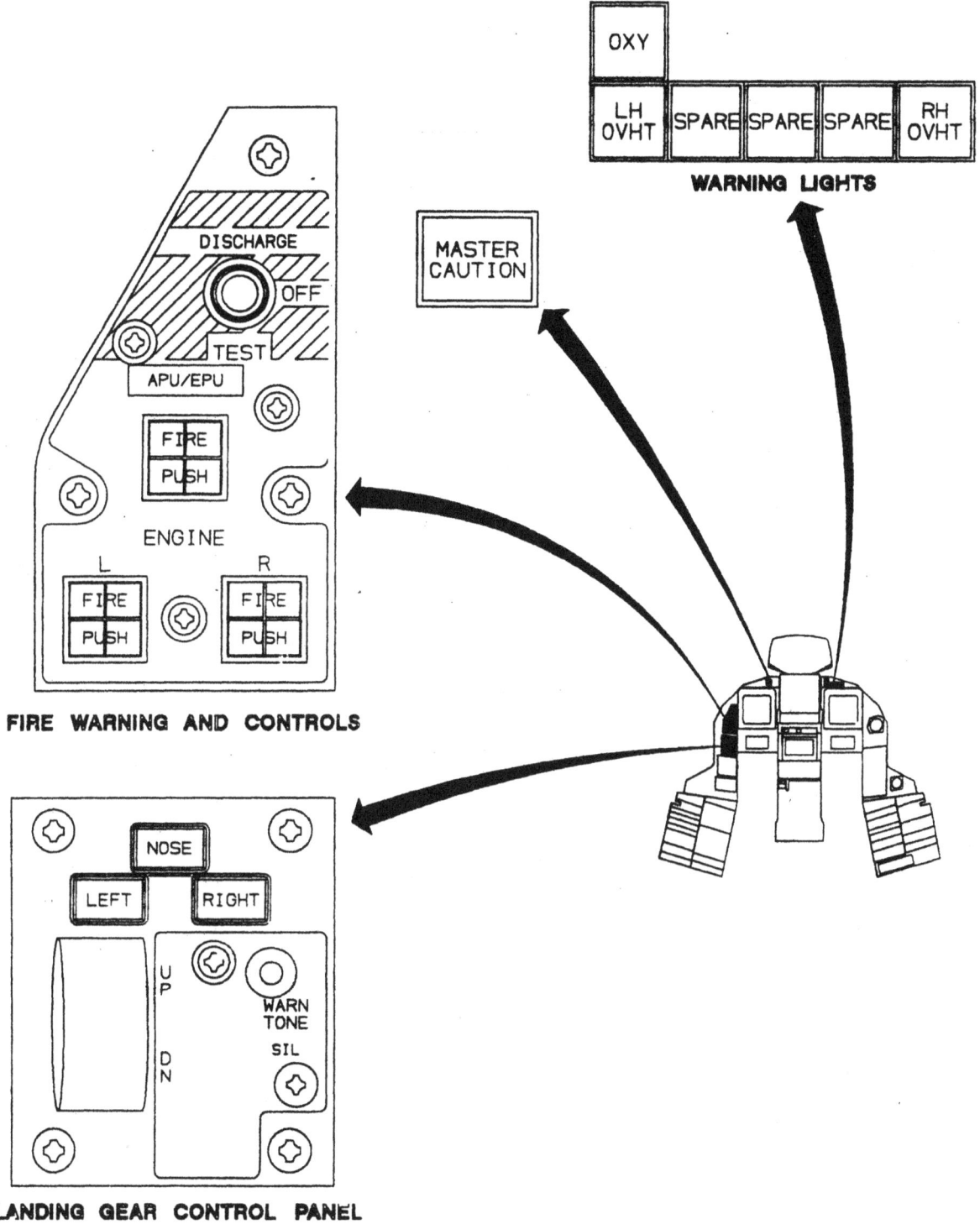

| OXY | | | | |
|---|---|---|---|---|
| LH OVHT | SPARE | SPARE | SPARE | RH OVHT |

**WARNING LIGHTS**

MASTER CAUTION

DISCHARGE

OFF

TEST

APU/EPU

FIRE PUSH

ENGINE

L

R

FIRE PUSH

FIRE PUSH

**FIRE WARNING AND CONTROLS**

NOSE

LEFT    RIGHT

U P

D N

WARN TONE

SIL

**LANDING GEAR CONTROL PANEL**

C29-005

*Figure 3-1*

EMERGENCY PROCEDURES
# GROUND

## ENGINE MALFUNCTION DURING START

If any abnormal engine indication, hung start, over-heat, bleed air leak, engine caution occurs during start or engine auto-accelerates beyond idle, shutoff engine and investigate. If starting EGT is exceeded, motoring the engine assists engine cool down and minimizes engine damage, see HOT START.

1. Throttle -- OFF.
2. Engine -- Motor (if necessary).
3. Engine Crank Switch -- OFF.

## ENGINE AUTO-ACCELERATION

If the engine auto-accelerates above idle RPM with the throttle in idle detent, shut down the engine and investigate. Loss of engine/VMS communication could require the use of the engine fire light button to shutoff fuel to the engine.

1. Throttle -- OFF.

**If engine fails to shutdown**

2. Engine Fire Button -- Press.

## HOT START

Hot start is a sub idle stall and exhaust gas temperature (EGT) exceeds engine limits. Possible hot start is indicated when RPM fails to increase to idle (or decreases) while EGT continues to increase.

1. Throttles -- OFF.
2. Engine -- Motor (if possible).
3. Engine Crank Switch -- OFF.

## ENGINE/AMAD BAY FIRE DURING START

Engine or AMAD bay fire is recognized by illumination of the fire warning light, voice *WARNING ENGINE FIRE LEFT (RIGHT)*, smoke and fumes, or notification by the ground crew. An engine fire lights the top of the fire light and an AMAD bay fire lights the bottom of the fire light. Fire detection system could take 60 seconds to reset after fire is extinguished.

If the top of the fire light comes on, activating the fire extinguishing system arms both fire extinguisher bottles and shuts off fuel and ram air to the engine. If only the bottom of the fire light comes on, only one fire extinguishing bottle is armed.

1. Throttles -- OFF.
2. Engine Fire Button -- Press.
   Lift metal guard up and press button. Arms **extinguisher agent discharge valve** and shuts off fuel to the engine.
3. Extinguisher Discharge Switch -- DIS-CHARGE.
   Agent discharge is immediate. If fire warning light goes out, test fire warning system.
4. Engine Crank Switch -- OFF.

## NOSEWHEEL STEERING CONTROL FAILURE

Nosewheel steering failure is indicated by the caution NWS, loss of steering, or uncommanded steering. The nosewheel steering system is deactivated while the paddle switch is held allowing the nosewheel to free caster. Use flight controls and differential braking to maintain directional control.

1. Paddle Switch -- Press and hold.
2. Use differential braking.

## BRAKE SYSTEM MALFUNCTIONS

A brake system malfunction is indicated by loss of braking or abnormal braking resulting from brake-by-wire (BBW) control failure, antiskid failure, or loss of hydraulic pressure to the brakes.

### BBW CONTROL FAILURE

BBW control failure is caused by the loss of the VMS communication between the brake pedal sensors and the antiskid control unit, or antiskid system failure. If BBW failure is detected, the brake control switch is electrically released to the mechanical (MECH) position. Antiskid protection is provided when braking in the mechanical control mode.

### ANTISKID FAILURE

Antiskid failure is indicated with lack of braking response and the caution ANTISKID.

## BRAKE HYDRAULIC FAILURE

If an abnormal braking condition is experienced without an indication of either BBW or antiskid failure, suspect brake hydraulic failure. Selecting the position BRAKE with the emergency brake switch provides an alternate source of hydraulic pressure through independent lines to the brakes.

### ⌐ CAUTION ¬

Brake control switch changes with brake pedal pressure applied could result in excessive braking, loss of braking, or tire failure. If abnormal braking or loss of braking is experienced, release brake pedal pressure prior to making brake control switch changes.

1. Brake Switch -- MECH.

**If caution ANTISKID does not rescind, or braking is not restored**

2. Antiskid Switch -- Off.

3. Brakes -- Apply light to moderate braking (conditions permitting).

**If braking is not restored**

4. Brake/NWS Emergency Control Switch -- BRAKE.

## EMERGENCY GROUND EGRESS

The boarding ladder provides the safest means to exit the aircraft.

1. Throttles -- OFF.

2. Ejection Handle Safety Lever -- Up.

3. Canopy -- OPEN.

4. Battery Switches -- OFF.

5. Lap Belt, Survival Kit Straps, Shoulder Harness -- Release.

6. Abandon Aircraft.

## EMERGENCY ENTRANCE

If access to the nose wheel well canopy switch or mechanical crank fitting is blocked and time is critical, the fire department gain entry to the cockpit by cutting the canopy (figure 3-2).

## CANOPY EMERGENCY OPENING

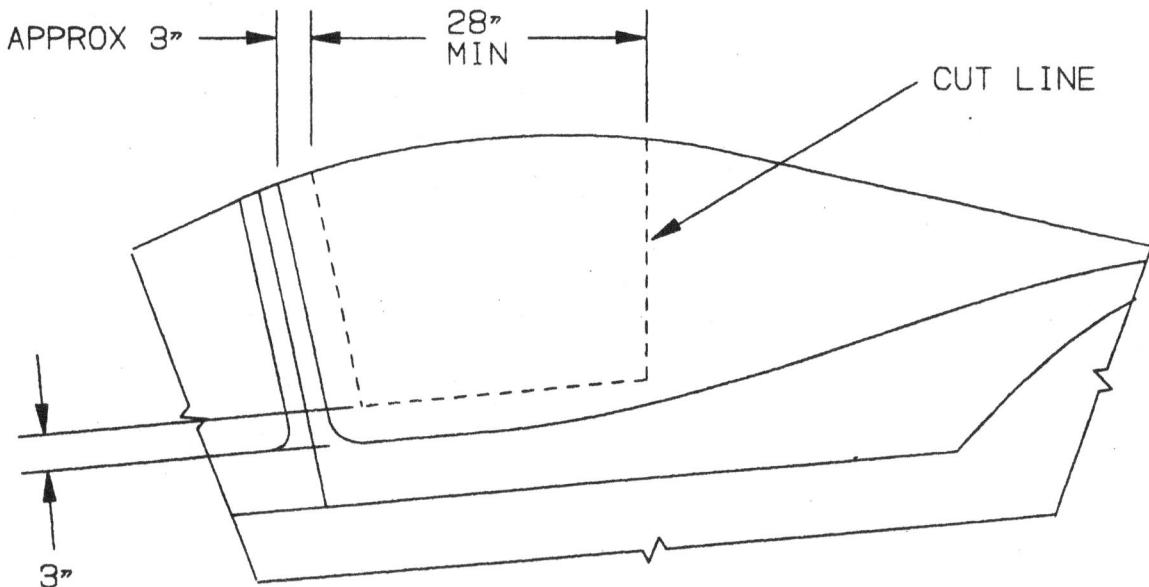

C23-011

**Figure 3-2**

1. External Canopy Switch -- OPEN (hold).
   The external mounted (nose wheel well) canopy switch must be held to the open position until the canopy is open.

2. Throttles -- OFF.

3. Ejection Seat Safety Lever -- Up.
   Safety lever is located behind of the left ejection handle, marked with solid yellow paint. Rotate handle forward and until locked up.

**WARNING**

Do not actuate either ejection seat firing handle. Rotate the ejection seat safety lever forward and up against the left ejection handle to safety the seat.

4. Lap Belt -- Release.

**WARNING**

Do not use the restraint emergency release handle for emergency rescue because of probable chute entanglement during pilot extraction. Manual release of lap belt, survival kit straps, and shoulder harness is recommended.

5. Survival Kit Straps -- Release.

6. Shoulder Harness -- Release.

# EMERGENCY PROCEDURES
# TAKEOFF

## ABORT

Use full aft stick to maximize aerodynamic drag and increase braking effectiveness. See HOT BRAKES following a high speed abort.

1. Throttles -- IDLE.

2. Control Stick -- Full aft.

3. Speed Brake Switch -- Forward.

## HOT BRAKES

After high speed abort, landing immediately after takeoff or anytime hot brakes are suspected, see Brake Energy Limitations chart (figure 3-3), to determine brake energy zone. With gross weight and braking speed, enter chart to determine if brake energy is in the danger, restricted, caution, or normal limitations zone.

### DANGER ZONE

Expect wheel/brake destruction and possible axle damage.

- Maintain forward motion at slow speed using light braking, only as necessary.
- Proceed to nearest parking area clear of aircraft and personnel without stopping. Use brakes to hold aircraft stationary until engine shut down.
- Request fire fighting equipment.
- Shut down engines.
- The area within 50 feet of either main landing gear brake is unsafe.
- Emergency egress forward of the aircraft.

### RESTRICTED ZONE

Expect wheel thermal release plugs to blow.

- Taxi below 25 knots ground speed to nearest parking area. Use minimum braking to hold aircraft stationary until engines are shut down.
- Advise maintenance personnel of brake condition. The area within 50 feet of either main landing gear brake is unsafe for one hour or until the wheel thermal release plugs have blown.

### CAUTION ZONE

Thermal release plugs could blow. Removal and inspection of wheel and tire assemblies is required.

- Taxi to designated parking area and shut down engines, if practical.
- Advise maintenance personnel of brake condition. The area within 50 feet of either main landing gear brake should be reguarded as unsafe for 20 minutes or until the wheel thermal release plugs have blown.

### NORMAL ZONE

No limitations.

## AFTERBURNER FAILURE DURING TAKEOFF

The engine has an automatic afterburner recycle capability. If after recycle the afterburner fails to relight the caution ENG 1ST appears and throttle should be retarded to MIL. See engine manufacturer's performance specifications

**If caution ENG 1ST/ENG 2ND appears**

1. Throttle -- MIL.

## ENGINE/AMAD BAY FIRE DURING TAKEOFF -- ABORT

1. Abort.

2. Throttle Affected Engine -- OFF.

3. Engine Fire Button -- Press.
   Lift metal guard up and press button. Arms extinguisher agent discharge valve and shuts off fuel and ram air to the engine.

4. Extinguisher Discharge Switch -- DISCHARGE.
   If fire warning light goes out, test fire warning system.

## ENGINE/AMAD BAY FIRE DURING TAKEOFF -- TAKEOFF CONTINUED

If a fire occurs after refusal speed and the decision is to continue the takeoff, climb to safe ejection altitude, and accomplish ENGINE/AMAD BAY FIRE IN FLIGHT.

1. Throttles -- As required.

## BRAKE ENERGY LIMITATIONS

### CONDITIONS:

- EAFB HOT DAY
- IDLE
- THREE POINT ATTITUDE
- HI DRAG FLAPS
  (SPEEDBRAKES EXTENDED)

- MAX AFT STICK
- MAXIMUM CONTINUOUS BRAKING
- TAXI ENERGY INCLUDED

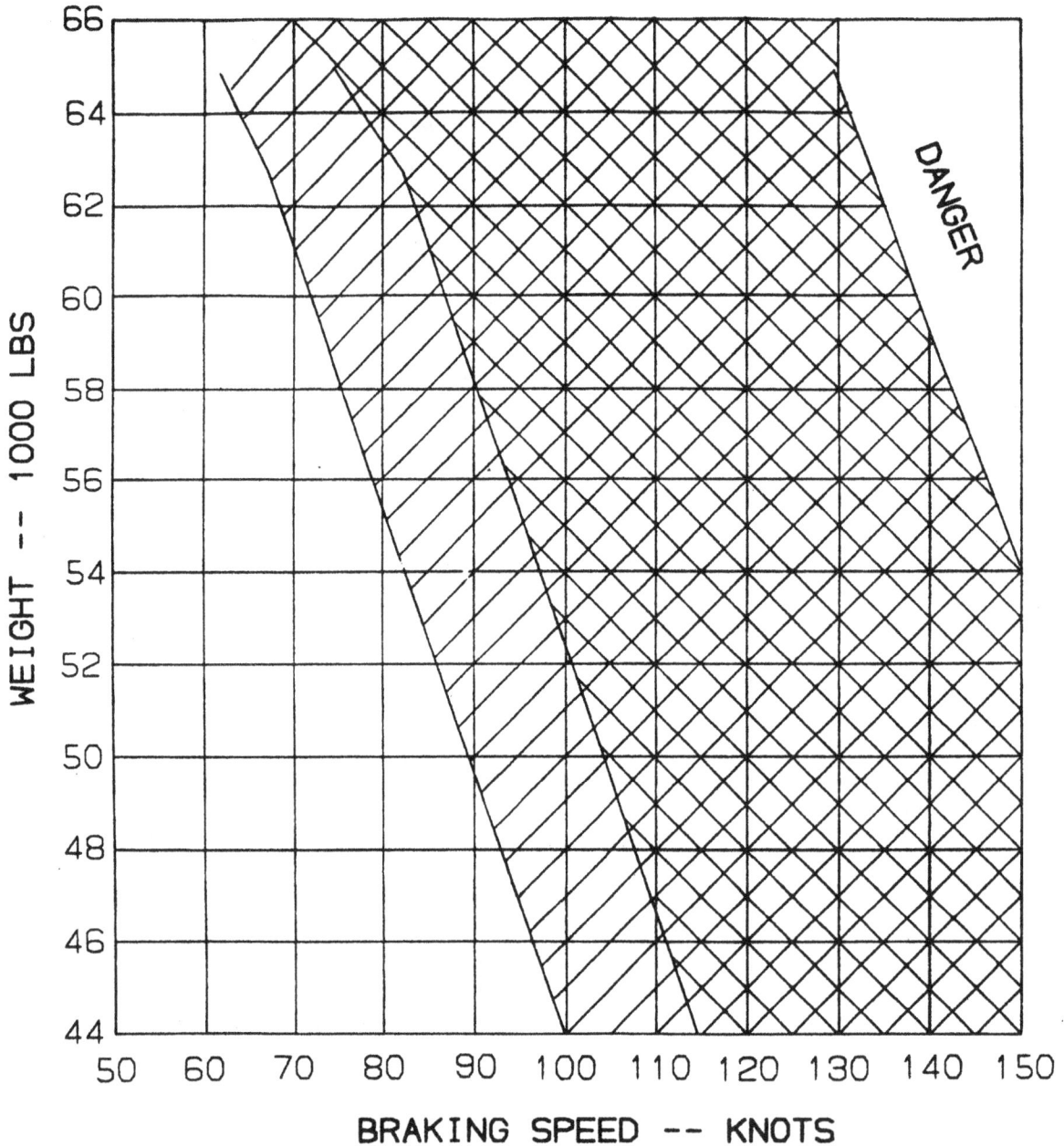

WEIGHT -- 1000 LBS

BRAKING SPEED -- KNOTS

DANGER

⬛ CAUTION   ❎ RESTRICTED

C23-013

*Figure 3-3*

2. Throttle Affected Engine -- IDLE (conditions permitting).
   If fire warning light goes out, test fire warning system.

**If fire warning light remains on**

3. Throttle Affected Engine -- OFF.

4. Engine Fire Button -- Press.

5. Extinguisher Discharge Switch -- DISCHARGE.

**If fire persists**

6. Eject.

# ENGINE FAILURE DURING TAKEOFF

Engine failure is recognized by loss of power, asymmetric thrust response, an engine caution/warning and master caution light. If takeoff is continued, climb to safe ejection altitude and accomplish inflight emergency procedures. Rotation and lift-off distance increase significantly for single engine takeoff. Single engine climb speed and single engine handling qualities.

In the event of an engine failure after nosewheel liftoff, neutralize rudders prior to lowering nosewheel to avoid directional control problems or damage to the nosegear.

**ENGINE FAILURE DURING TAKEOFF -- ABORT**

1. Abort.

### CAUTION

Neutralize rudder pedals prior to nosewheel touchdown if abort is after rotation. Nosewheel structural damage could occur if sideloads are excessive.

2. Throttle Affected Engine -- OFF.

**ENGINE FAILURE DURING TAKEOFF -- TAKEOFF CONTINUED**

Continued takeoff with an engine failure while still on the ground is not recommended. With an engine failure, flight control authority to rotate and maintain zero directional drift is insufficient. Techniques have been successful in performing the maneuver in the simulator, but continued takeoff is not recommended.

- Throttles -- As required.
- Maintain directional control with rudder pedals.
- Accelerate to the highest possible speed in the runway available, not to exceed tire limit speed.
- Release pedals and rotate with full aft stick.

# DOUBLE GENERATOR FAILURE DURING TAKEOFF

Double generator failure on takeoff is indicated by blanking of HUD, MPCD, UFC, and loss of engine thrust as the fuel flow is cutback until the emergency generator comes on line.

1. Abort.

**If takeoff continued**

2. Throttles -- MIL.

# GEAR FAILS TO RETRACT

If the gear fails to retract, put the gear handle down and do not attempt to retract the gear again.

1. Gear Handle -- DN.
   Remain below 250 KCAS. If gear indicates down and locked, reduce gross weight and land. If gear does not indicate down and locked, see ABNORMAL GEAR CONFIGURATION.

# TIRE FAILURE DURING TAKEOFF

1. Abort.

**If main gear tire fails**

2. Antiskid Switch -- OFF.

**If takeoff continued**

1. Gear -- Do not retract.
   Do not exceed landing gear speed, see LANDING WITH TIRE FAILURE.

## EMERGENCY PROCEDURES
# INFLIGHT

## BLEED AIR LEAK

Bleed air leaks are recognized by illumination of the caution messages L BLD LEAK/R BLD LEAK. Advisories L BLEED and R BLEED indicate the primary bleed air valves have closed. A combination of these cautions and the advisories indicate where the failure has occurred and what action should be taken to correct the situation.

A caution L BLD LEAK or R BLD LEAK indicates a leak in the respective engine bleed air line. Check for a corresponding advisory L(R) BLEED to indicate bleed air valve has closed. If the advisory does not appear, close the bleed air valve by selecting the unaffected side with the bleed air control and retard throttle on affected side to idle. If valve does not close, shut down the affected engine to stop the bleed air leak.

Both cautions L BLD LEAK and R BLD LEAK, with both advisories L BLEED and R BLEED, indicate a bleed air leak between secondary regulator and the primary heat exchanger, and the bleed air valves have closed. The forward avionics bay requires ram air cooling.

Both cautions L BLD LEAK and R BLD LEAK, and no advisories (L BLEED and R BLEED) indicates a bleed air leak in the forward part of the aircraft and requires ram air cooling to avionics bay.

An undetected bleed air leak between the primary heat exchanger and the air cycle machine could be indicated by loss of performance and the caution ECS AIR HOT. If the caution ECS AIR HOT is accompanied with a loss of engine performance, turn ECS bleed air control to OFF.

## BLEED AIR LEAK -- SINGLE

A caution L BLD LEAK or R BLD LEAK indicates a single bleed air leak.

1. ECS Bleed Air Control -- Select unaffected side.
   Check corresponding bleed air advisory (L BLEED or R BLEED) comes on to confirm valve closed.

2. Throttle Affected Engine -- IDLE.
   Advisory L(R) BLEED indicates valve closed and bleed leak has been isolated. Caution L(R) BLD LEAK rescinds.

**If caution L BLD LEAK or R BLD LEAK remains on**

3. Throttle Affected Engine -- OFF (if practical).

4. Land as soon as practical.

## BLEED AIR LEAK -- DOUBLE

**Both L BLEED and R BLEED and no L BLD LEAK or R BLD LEAK**

1. ECS Bleed Air Control -- Cycle.
   Cycle control to OFF and back to BOTH.

**Both L BLD LEAK and R BLD LEAK, or**

**Both L BLEED and R BLEED remain lit**

2. Oxygen -- As required.
   If failure occurs above 10,000 feet MSL. emgency oxygen should be activated without delay.

   a. Emergency Oxygen -- Activate.

   b. Oxygen Power Switch -- POWER OFF.

   c. Oxygen Supply Switch -- OFF.

3. ECS Mode Switch -- OFF/RAM.

**When below 10,000 feet MSL**

1. Cabin Pressure Switch -- RAM/DUMP.

2. ECS Bleed Air Control -- OFF.

3. Land as soon as possible.

## ECS OVERTEMPERATURE

Overtemperature condition in the avionics equipment bay is recognized by the caution ECS AIR HOT. The condition can be caused by insufficient air flow or hot air flow to avionics equipment and vehicle management computer. If cooling air is not provided within two minutes, equipment failures could occur. The first response should be to cycle the ECS mode switch to manual.

The advisory L (R) BLEED could indicate a failed ed bleed air valve and cycling the bleed air control to OFF then BOTH could reestablish bleed air flow.

If ECS bleed air cooling is discontinued, optimum ram air cooling obtained at 10,000 feet MSL, is between 0.8 - 0.9 Mach.

1. ECS Mode -- MAN.
   Provides maximum airflow to cockpit and avionics bay.

2. ECS Bleed Air Control -- OFF, then BOTH.

3. Either Throttle -- Advance by 4% RPM.

4. Oxygen -- As required.
   If failure occurs above 10,000 feet MSL, emergency oxygen should be activated without delay.

   a. Emergency Oxygen -- Activate.

   b. Oxygen Power Switch -- POWER OFF.

   c. Oxygen Supply Switch -- OFF.

5. Airspeed -- Maintain subsonic.

6. ECS Mode Switch -- OFF/RAM.

7. Altitude -- Descend below 10,000 feet MSL.

If light remains on after 2 minutes

8. Cabin Pressure Switch -- RAM/DUMP.
   The emergency ram air scoop opens providing ram air cooling to the cockpit and avionics bay. ECS and oxygen generating system are turned off.

9. Nonessential Electrical Equipment -- Off.

10. Land as soon as practical.
    Vehicle management system has an estimated 30 minute overtemperature life-expectancy.

**If below 10,000 feet MSL**

4. Cabin Pressure Switch -- RAM/DUMP.

5. Airspeed -- Maintain subsonic.

If light remains on

6. Nonessential Electrical Equipment -- Off.

7. Land as as soon as practical.
   Vehicle management system has an estimated 30 minute overtemperature life-expectancy.

---

!CAUTION!

Turn off all avionics equipment as soon as possible after landing to avoid equipment damage.

## OXYGEN SYSTEM FAILURE

The warning light OXY and the voice warning *THREAT WARNING* indicate oxygen system failure. Activating the emergency oxygen bottle supplies unregulated 100% oxygen. Turning off the oxygen regulator prevents any dilution of the emergency oxygen supply.

Emergency oxygen duration is approximately 5-15 minutes. Altitudes above 38,000 feet MSL and mask leakage cause shorter durations.

**Above 10,000 feet MSL**

1. Emergency Oxygen Bottle Handle -- Activate.
   Provides 5-15 minutes of oxygen.

2. Oxygen Power Switch -- POWER OFF.

3. Oxygen Supply Switch -- OFF.

4. Altitude -- Maintain below 10,000 feet MSL.

## EXTREME COCKPIT TEMPERATURE

If the cockpit temperature is not controllable in automatic or manual, select RAM/DUMP to stop incoming conditioned hot air.

1. ECS Mode Switch -- Manual.

2. Temperature Control Knob -- Full cold.

3. Airspeed -- Maintain subsonic.

4. Oxygen -- As required.
   If failure occurs above 10,000 feet MSL, emergency oxygen should be activated without delay.

   a. Emergency Oxygen -- Activate.

   b. Oxygen Power Switch -- POWER OFF.

   c. Oxygen Supply Switch -- OFF.

5. Cabin Pressure Switch -- RAM/DUMP.
   The emergency ram air scoop opens providing ram air cooling to the cockpit and avionics bay. ECS and oxygen generating system are turned off.

6. Land as soon as possible.
   VMS without sufficient cooling air is approximately 30 minutes.

# AIR DATA LOSS -- TOTAL

The caution AIR DATA indicates a degraded mode of air data information but valid control laws. The caution FX GAINS indicates a total loss of air data information. The controls laws use the fixed air data based on airspeed and altitude at the time of failure. The VMS uses flight control laws based on flight conditions at high subsonic range, if supersonic. If flight control is degraded, fixed gain could provide better control feel.

### AIR DATA LOSS ABOVE 350 KCAS

1. Fixed Gain -- HI A/S.
   Selects fixed flight control gain for high subsonic flight. Adjust airspeed as necessary for stick feel force.

2. Airspeed -- Slow below 1.0 Mach.

3. See Air Data Loss Below 350 KCAS.

### AIR DATA LOSS BELOW 350 KCAS

1. Fixed Gain Switch -- LO A S.

**Prior to landing**

2. Airspeed -- 220 KCAS.

3. Landing Gear Down -- 220 KCAS.

4. Controllability -- Check.
   Slow to minimum control speed or minimum approach speed which ever is higher.

5. Land as soon as practical.

# FLIGHT CONTROL MALFUNCTIONS

In addition to abnormal response, degraded flight controls are indicated by vehicle management system cautions. The caution RESET VMS accompanies the cautions FCS 1ST, FCS 2ND, and FCS 3RD. Cease heavy maneuvering. Attempt vehicle management system reset. If caution FCS 1ST can be reset, normal flight can be resumed. The mission should be discontinued if VMS does not reset or the cautions FCS 2ND/FCS 3RD appear.

1. Paddle Switch -- Squeeze and release, if flight test gains are enabled.
   The paddle switch reinstates normal control laws if the flight control malfunction is a result of flight test gain control.

2. VMS/ENG -- RESET.

If Caution FCS 2ND/FCS 3RD appear, or caution FCS 1ST did not reset.

3. Airspeed -- Maintain 200-300 KCAS.
   If airspeed is not reliable, fly approximately 4-6 degrees AOA.

4. VMS/ENG -- RESET.

5. Controllability -- Check.
   See CONTROLLABILITY CHECK.

6. Land As Soon As Practical.

# FLAP MALFUNCTIONS

A flap malfunction is indicated by the caution FLAPS or an uncommanded rolling/yawing moment during landing gear extension/retraction.

### FLAPS FAILURE

The cause of the caution FLAPS, could be the loss of auto flaps. Flaps could remain in the last position attained at the time of failure. Use emergency flaps or make no flap landing.

### FLAP ASYMMETRY

If lateral rolling and yawing occurs during operation of the wing flaps, or the caution FLAPS appears. suspect an asymmetrical flap condition. If a leading edge flap malfunction can be confirmed, do not exceed 15 degrees AOA and 45 degrees of bank.

1. VMS/ENG -- RESET.

2. Trim -- As required.

**If insufficient roll trim**

3. Flap Emergency Switch -- EMERG UP.
   Plan no flap landing.

**If roll continues**

3. Emergency Flap Switch -- EMERG DN.
   Conditions permitting, slow aircraft to below 250 KCAS.

4. Controllability -- Check.
   See CONTROLLABILITY CHECK.

# OUT OF CONTROL RECOVERY

1. Controls -- Neutralize.
   Neutral position nulls all pilot commanded flight control inputs.

2. Paddle Switch -- Press and release.
   The paddle switch reinstates normal control laws if the flight control malfunction is a result of flight test gain control.

3. Throttles -- As required.
See engine manufacturer's performance specifications. An out of control condition can be aggravated by asymmetrical thrust.

**If not recovered and below 200 KCAS**

4. Flight Control Override Switch -- ON.

5. Paddle Switch -- Press and hold.
When flight control override switch is on and paddle switch is held, maximum flight control displacement is provided.

6. Stick -- As required.

**After recovery**

7. Paddle Switch -- Release.

8. Flight Control Override Switch -- OFF.

**If not recovered by 10,000 feet AGL**

7. Eject.

## SPIN RECOVERY

### ERECT SPIN RECOVERY

Apply out of control recovery procedures before applying spin recovery procedures.

1. Paddle switch -- Press and hold.

2. Stick -- Full forward.

3. Ailerons -- Full with spin.

4. Rudder Pedals -- Full against spin.

**After recovery**

5. Paddle Switch -- Release.

6. Flight Control Override Switch -- OFF.

**If not recovered by 10,000 feet AGL**

5. Eject.

### INVERTED SPIN RECOVERY

Apply out of control recovery procedures before applying inverted spin recovery procedures.

1. Paddle switch -- Press and hold.

2. Stick -- Full aft.

3. Ailerons -- Neutral.

4. Rudder Pedals -- Full against spin.

**After recovery**

5. Paddle Switch -- Release.

6. Flight Control Override Switch -- OFF.

**If not recovered by 10,000 feet AGL**

5. Eject.

## ENGINE MALFUNCTIONS

In addition to abnormal engine performance, the cautions L(R) ENGINE, L(R) ENG 2ND, and L(R) ENG 1ST indicate an engine malfunction or engine performance limit. The caution RESET VMS could appear with the engine cautions. Check standby flight instrument panel for additional indications.

If engine exceeds operating limitations, adjust power to maintain normal limits and land as soon as practical. Shutdown engine if normal limits can not be maintained.

If engine stalls, rolls back, or shuts down, press engine reset. Restart engine.

If engines do not respond to throttle cutoff, the engine fire light buttons can be used as an alternative means to shutoff fuel to the engines.

A windmilling engine produces insufficient oil pressure for lubrication. See engine manufacturer's performance specifications.

1. Throttle Affected Engine -- IDLE.

2. VMS/ENG -- RESET.

3. Standby Flight Display 1 -- Check.

4. Throttle Affected Engine -- As required.
See engine manufacturer's performance specifications.

## ENGINE OIL PRESSURE MALFUNCTION

Engine oil system malfunctions include zero, low, and high oil pressure. Zero oil pressure causes a master caution light and a caution L(R) ENGINE. Low oil pressure causes a master caution light and a caution L(R) ENG 1ST. For high oil pressure see engine manufacturer's performance specifications. Standby flight display one and the MPCD engine display provide oil pressure information.

Inverted flight could cause temporary oil system malfunction indications. If oil pressure returns to normal after inverted flight, continue normal operations.

If oil pressure stays at zero or is abnormally high and the other engine is operating normally, the malfunctioning engine should be shut down to minimize damage. If oil pressure is low and appreciable time remains during recovery, consider engine shutdown during recovery and restart prior to landing. If abnormal oil pressure is accompanied by engine vibration, shut the engine down.

1. Throttle - IDLE.

2. VMS/ENG -- RESET.

3. Standby Flight Display 1 - Check.

**If oil pressure is zero or high**

4. Throttle Affected Engine - OFF:

5. Land as soon as practical.

**If oil pressure is low**

4. Throttle Affected Engine - As Required. Operate engine at idle during recovery. Minimize G forces.

5. Land as soon as practical.

## ENGINE STALL

The engine control unit detects engine stall based on rate of change of engine burner pressure. Automatic stall recovery actions include afterburner cancellation, engine ignition, and limited fuel flow during engine recovery and reacceleration. If automatic stall recovery fails, a master caution light could come on accompanied by rising EGT and decreasing RPM.

1. Throttle -- Out of AB.

**If engine does not recover**

2. Throttle -- IDLE.

**If stall does not clear**

3. Throttle -- OFF.

4. Engine -- Airstart.
If restart is successful, leave engine at idle unless additional thrust is required for safe recovery.

## DOUBLE ENGINE FLAMEOUT

Emergency power unit start up is automatic. Advisories EPU HYD, HYDRAZINE, and EPU GEN indicate emergency power unit has started. See EMERGENCY POWER OPERATION.

If the EPU is activated, see LANDING WITH EPU ACTIVATED.

1. Throttles -- Idle.

2. Airspeed -- As required.

3. EPU Selector -- ON.
Push selector in and rotate to ON.

4. Engines -- Airstart.

**Above 10,000 feet MSL**

5. Emergency Oxygen Bottle Handle -- Activate. Provides 5-15 minutes of oxygen.

6. Oxygen Power Switch -- POWER OFF.

7. Oxygen Supply Switch -- OFF.

## EMERGENCY POWER OPERATION

The emergency power unit (EPU) comes on automatically for double engine flameout, double generator failure, or loss of the utility hydraulic system and either flight control hydraulic system. If the EPU fails to come on automatically, move the EPU control switch to ON.

Emergency power unit operation is indicated by the advisories EPU GEN, HYDRAZINE, and EPU HYD, and the decrease in hydrazine quantity on the standby flight display 2.

The hydrazine supply provides enough power to run the EPU emergency generator for approximately 15 minutes and approximately 9 minutes with the emergency hydraulic pump providing hydraulic power. If the emergency generator fails approximately 2 minutes of battery power is available.

If the EPU is activated, see LANDING WITH EPU ACTIVATED.

1. EPU Selector -- ON.

## ENGINE AIRSTART

Airstarts using bleed air, spooldown, or windmilling engine can be accomplished within the airstart envelope. See engine manufacturer's performance specifications.

1. Airspeed -- As required.
See engine manufacturer's performance specifications.

2. Throttle -- OFF.

3. VMS/ENG -- RESET.

4. Throttle -- IDLE.

**If no start by 20 % RPM**

5. Operating Engine -- RPM as required.
See engine manufacturer's performance specifications.

6. Engine Crank Switch -- L or R (as required).
Place switch to select the failed engine.

## ENGINE/AMAD BAY FIRE IN FLIGHT

If the engine fire light (upper half) comes on, activating the fire extinguishing system arms both fire extinguisher bottles and shuts off fuel and ram air to the engine. If the AMAD fire light (lower half) is on, only one fire extinguishing bottle is armed.

If an explosion or catastrophic engine failure occurs, do not delay engine shutdown. After the fire light has been pressed, do not press again. If the fire extinguishing agent is used, do not restart engine unless absolutely necessary.

> | WARNING |
>
> Pressing the engine fire button a second time reopens main fuel shutoff valve and allows fuel to flow to the engine. Possible fire or engine explosion could result.

1. Throttle Affected Engine -- IDLE.

2. Standby Flight Display 1 -- Check.

**If fire light remains on for 30 seconds or fire is confirmed**

3. Throttle Affected Engine -- OFF.

4. Engine Fire Button -- Press.
Lift metal guard and press button.

5. Extinguisher Discharge Switch -- DISCHARGE.

**If fire warning light goes out**

6. Fire Warning System -- Test.

7. Bleed Air Control -- Cycle to OFF, then as required.

**If fire persists**

6. Eject.

## EPU FIRE IN FLIGHT

1. APU/EPU Fire Button -- Press.
[2] APU fuel is shutoff, one fire extinguishing agent bottle is armed, and a 10 second delay is started (allows APU to spool down).

**Wait 10 seconds**

2. Extinguisher Discharge Switch -- DISCHARGE.
Time delay prevents discharge 10 seconds until after fuel shut off. If fire warning light goes out, test fire warning system.

**If fire light remains on**

3. APU/EPU Fire Button -- Press twice.
Arms second fire extinguisher bottle.

**Wait 10 seconds**

4. Extinguisher Discharge Switch -- DISCHARGE.
Time delay prevents discharge for 10 seconds. If fire warning light goes out, test fire warning system.

**If fire persists**

5. Eject.

## FUEL MANAGEMENT COMPUTER FAILURE

The caution FUEL FAIL indicates failure of the fuel management computer or a fuel gauging probe failure. The caution CG comes on.

1. Fuel Display -- Check.
If invalid timer appears (bottom right corner of display), the fuel management computer has failed. If EST is shown by a fuel tank, a fuel probe has failed. If INV is shown by a fuel tank, tank quantity reading is invalid.

**If invalid timer appears on fuel display**

2. AOA -- 12 degrees, maximum.

3. Land as soon as practical.

## FUEL TANK PRESSURIZATION SYSTEM FAILURE

Low tank pressure is indicated by the caution TANK PRESS. Maintain below 17,000 feet MSL (if practical) to prevent fuel pump cavitation.

1. Throttles -- Out of AB.

2. Altitude -- Maintain below 17,000 feet MSL.

3. Land as soon as practical.

## FUEL ASYMMETRY

Indication of a fuel asymmetry could be:

- Caution FUEL ASYM (wing transfer pump failed).

- Caution CG (fuselage transfer pump failed).

The caution FUEL ASYM indicates a 500 pounds fuel imbalance between the left and right wing tanks, and could indicate the failure of a wing transfer pump on one side.

A caution CG could be an early indication of fuel transfer failure. Monitor CG and fuel tank quantities.

1. Throttles -- Out of AB.
   If a low fuel state is suspected in the main transfer tank(s), retard throttles to minimum practical setting.

2. Maneuvering -- Discontinue.

3. Manual Fuel Control Display -- Select.
   Press button labeled CONT on fuel display. Manual fuel control display appears. Check tank quantities.

### CAUTION

Do not deselect the manual fuel display for more than 30 seconds. The fuel system returns to automatic mode within 30 seconds after deselecting the manual fuel display.

3. Fuel Diverter Valves -- Verify feed position.
   Press button labeled LDIV and button labeled RDIV until the letter F appears.

4. Wing Transfer Pumps -- Off.
   Press the button labeled LWP and button labeled RWP. Check word OFF appears beside label.

5. Land as soon as practical.

## FUEL LOW

Indication of a fuel low could be:

- Premature caution FUEL LOW,
- Voice *WARNING, FUEL LOW* ,

Low fuel status is indicated by the caution FUEL LOW, accompanied by the voice *WARNING, FUEL LOW* . The low fuel caution is activated by low fuel level in the engine feed tanks.

The caution FUEL LOW indicates the engine feed tanks are not being replenished from the wing or fuselage transfer tanks. Any delay in reducing thrust could deplete feed tank fuel causing engine flameout.

1. Throttles -- Out of AB.

2. Fuel Status Display -- Select.

3. Fuel Quantity -- Verify fuel in feed tanks.

#### If all fuel is not in feed tanks

4. Paddle Switch -- Squeeze and release.
   Returns fuel control to automatic.

5. Land as soon as possible.

#### If one fuel tank is low and fuel leak confirmed

6. Throttle Affected Engine -- OFF.

7. Engine Fire Button (affected side) -- Press.
   The engine fire warning light closes the fuel shutoff valve isolating leaks downstream of the valve.

#### If fuel caution CG is on

8. AOA -- Limit to 12 degrees.

## FUEL BOOST PUMP FAILURE

Failure of one boost pump causes the crossfeed valve to open automatically. One boost pump is sufficient to provide fuel for both engines at throttle settings up to MIL power.

Double generator failure could cause both boost pumps to stop until the EPU provides emergency power. Normal engine operation cannot be sustained with failure of both fuel boost pumps. The DC fuel pumps can supply 5000 lb/hr/pump fuel flow.

1. Throttles -- Out of AB.

2. Fuel Status Display -- Select.
   Confirm boost pump failure by low fuel pressure.

#### NOTE

If one boost pump fails, limit throttles to MIL power range maximum. If both boost pumps fail, limit throttles to less than 5,000 lb/hr/engine fuel flow maximum.

3. Land as soon as practical.

## FUEL HOT

If the engine inlet fuel is above 115 degrees Centigrade, the caution FUEL HOT comes on. Fuel could vaporize and cause fuel pump cavitation and engine flameout is possible. A hot fuel condition also exists, if engine fuel temperature on standby flight display 1 is above engine fuel temperature limit or fuel feed tank temperature is above 60 degrees Centigrade.

1. Standby Flight Display 1 -- Select.
   Determine side affected by fuel temperature hot.

2. Throttle Affected Engine -- Set 5,000 pph fuel flow.
   If conditions permit, set 5,000 lbs/hr/engine fuel flow or more for fuel cooling.

**If caution persists/ fuel temperature confirmed hot**

3. Land As Soon As Practical.

## FUEL VENTING

Fuel venting could be caused by failure of the fuel level control valve in tanks 1 or 2. Fuel transfer is normally stopped when tanks 1 and 2 are full. If the fuel level control valve in either tank fails, fuel transfer continues and vents fuel overboard. Fuel venting can be confirmed by chase aircraft, if available.

Indication of venting could be a premature bingo fuel or fuel low cautions. If venting occurs during air refueling, discontinue refueling.

1. Manual Fuel Control Display -- Select.
   Monitor fuel level in feed tanks.

2. Left Feed Tank Refuel Shutoff Valve -- Close.
   Press button labeled LF. As feed tank fuel is used, manually open refuel control valve to refill engine feed tank.

**If venting continues**

3. Left Feed Tank Refuel Shutoff Valve -- Open.
   Press button labeled LF.

4. Right Feed Tank Refuel Shutoff Valve -- Close.
   Press button labeled RF. As feed tank fuel is used, manually open refuel control valve to refill engine feed tank.

5. Altitude -- Maintain below 25,000 feet MSL.

## SINGLE GENERATOR FAILURE

Caution L GEN or R GEN indicate generator failure. Essential flight electrical loads can be provided by one generator. Check the MPCD for associated cautions, AMAD and XFMR RECT.

Both left and right main AC bus 2 are inoperative with single generator failure.

1. EPU Selector -- AUTO.

2. Generator -- Reset.
   Cycle generator switch to RESET and back to ON. Generator may be reset as many times as required.

3. VMS/ENG -- RESET.

**If generator fails to reset**

4. Generator Switch -- OFF.

5. Land as soon as practical.

| AC Equipment Lost |
|---|
| ● Flight test equipment, |
| ● Fuel flow transmitter, |
| ● Programmable armament control (PACS), |
| ● Seat adjustment, |
| ● IFF Mode 4, |
| ● Weapons stations, |
| ● Wing fuel transfer (left and right). |

## DOUBLE GENERATOR FAILURE

If both generators fail, the emergency generator automatically starts and powers the essential buses. See EMERGENCY POWER OPERATION.

See LANDING WITH EPU ACTIVATED.

For affected electrical buses with double generator failure, see figure 3-4. To avoid engine stall or flameout, reduce thrust to idle until electrical power to the fuel boost pumps is regained.

If the electronic ADI display does not automatically come up on the right MPCD, use backup instruments and attempt to reset the emergency generator. If the right MPCD is still blank, attempt to reset by cycling the MPDP power switch to OFF, then ON.

Many other warnings, cautions, and advisories are shown, including voice messages and warning tones. Warning light OXY appears 10-20 seconds after failure if a generator cannot reset.

1. Throttles -- IDLE (immediately).
   Any delay in retarding throttles could result in dual engine flameout. Placing throttles to idle ensures rapid transfer of electrical supply to the emergency power unit. Advisory EPU GEN indicates emergency power unit 1 is operating.

# OPERATING ELECTRICAL BUSES

| BUSSES | EMERG GEN ON (15 MINS) | BATTERIES | |
|---|---|---|---|
| | | UTIL AND EMERG (2 MINS) | VMS 1 AND VMS 2 (2 MINS) |
| AC ESSENTIAL LEFT AND RIGHT | ON | — | — |
| AC MAIN 1 LEFT AND RIGHT | — | — | — |
| AC MAIN 2 LEFT AND RIGHT | — | — | — |
| DC ESSENTIAL LEFT AND RIGHT | ON | ON | — |
| VMS 1 AND VMS 2 | ON | ON | ON |
| DC MAIN LEFT | — | — | — |
| RIGHT | ON (IN FLIGHT) | — | — |

C23-012

**Figure 3-4**

!CAUTION!

Engine stall could occur with double generator failure. Idle **engine** thrust could help avoid engine stall by reducing engine fuel demands until the EPU can stablize electrical power to the fuel boost pumps.

2. EPU -- Check operating.
   If EPU fails to start, turn EPU on manually and reset the emergency generator, if required.

3. Throttles -- As required.

4. Oxygen -- As required.
   If failure occurs above 10,000 feet MSL, emergency oxygen should be activated without delay.

   a. Emergency Oxygen -- Activate.

   b. Oxygen Power Switch -- POWER OFF.

   c. Oxygen Supply Switch -- OFF.

5. Generators -- Reset.
   Cycle each generator switch to RESET and back to ON.

6. MPDP Power Switch -- Cycle OFF, ON if MPCD blank.
   Cycle switch momentarily (minimum of 0.5 seconds) to OFF, then ON, and release.

7. VMS/ENG -- RESET.

8. Land as soon as practical.

**If neither generator resets**

9. Land as soon as possible.
   Emergency generator power supply is approximately 15 minutes.

10. EPU Selector -- ON (just prior to landing).
    Just prior to landing cycle selector from AUTO to ON.

# EQUIPMENT LOST WITH DOUBLE GENERATOR FAILURE

| Additional AC Equipment Lost (EPU on) | DC Equipment Lost |
|---|---|
| • Jerkin controls | • Up front control (partial) |
| • Avionics interface unit 2 | • Tank 3 & 4 control |
| • HUD | • Antiskid |
| • INS | • Hot fuel valves |
| • Display processor (partial) | • UHF 2 |
| • Oxygen generation | • Fan bleed valve |
| • Pitot heat | • Seat adjust |
| • Flight test equipment | • TACAN |
| • Left boost pump | • ECS control |
| • Fuel transfer 1 | • Avionics Interface 2 |
| • Anticollision lights | • Oxygen generation control |
| • ECS | • Left MPCD |
| • Central computer | • Fuel valves |
| • TACAN | • Anticollision lights |
| • Aft ECS ram scoop | • **Armament loading isolation valve** |

## BATTERY FAILURE

The caution BATTERY indicates the emergency, utility, or one of the two flight control (VMS) batteries is failed.

1. Land as soon as practical.
   Mission control monitors electrical system and could provide information on affected battery.

## TRANSFORMER RECTIFIER FAILURE

If the caution XFRM RECT appears, use the avionics built-in test display to determine the transformer rectifiers lost.

1. Avionics BIT -- Select.
   Select avionics BIT menu and determine extent of loss. Mission control monitors electrical system and could provide assistance.

### ONE OR TWO TRANSFORMER RECTIFIERS LOST

Maximum load capacity is reached with the loss of two rectifiers.

1. Nonessential Electrical Equipment -- Off.
   • TACAN
   • IFF
   • UHF 2
   • Master Arm

2. Land as soon as practical.

### THREE TRANSFORMER RECTIFIERS LOST

Three transformer rectifier failures could result in the loss of some electrical equipment.

1. Non Essential Equipment -- Off.
   • TACAN
   • IFF
   • UHF 2
   • MASTER ARM
   • HUD
   • Left MPCD
   • Flight Test Data

2. Land as soon as possible.

## CENTRAL COMPUTER FAILURE

Central computer failure is indicated if the message STANDBY flashes on the left MPCD and the right MPCD changes to an electronic ADI. Accurate fuel CG information is lost to the VMS and the cautions TANK PRESS and FUEL ASYM appear. Fuel information for cautions FUEL FAIL, TANK PRESS and FUEL ASYM is inaccurate. Use fuel quantity from standby flight display.

1. Central Computer -- Reset.
   Press button labeled CC RESET located on the INS control panel.

### If in manual fuel control

2. Paddle Switch -- Squeeze.
   Returns fuel control to automatic.

### If central computer does not reset (left MPCD blank)

3. Central Computer Power Switch -- Cycle.
   Move switch through AUTO to OFF, then momentarily to ON.

4. MPDP Switch -- Cycle.
   Cycle multipurpose display processor power switch to OFF, then ON, and release.

5. Land as soon as practical.

## SMOKE OR FUMES IN THE COCKPIT

Consider all unidentified odors as toxic. Do not confuse condensation from the air conditioning system with smoke. The most probable source of smoke or fumes is from the engine or from residual oil in the ECS ducts. The smoke is blue grey, has a pungent odor, and could sting the eyes.

There are no fuel or hydraulic lines passing through or near the cockpit area. Possibility of cockpit fire is remote.

1. Oxygen Power Switch -- 100%.
Provides maximum oxygen enrichment at slight positive pressure.

**If smoke and fumes or odors persist**

2. Emergency Oxygen Bottle Handle -- Activate.
Provides 5-15 minutes of oxygen.

3. Oxygen Power Switch -- POWER OFF.

4. Oxygen Supply Switch -- OFF.

5. Cabin Pressure Switch -- RAM/DUMP.
The cockpit is unpressurized. Descend below 10,000 feet MSL before emergency oxygen supply is depleted.

**If electrical smoke confirmed**

6. Non-Essential Electrical Equipment -- Off.

7. Land as soon as possible.

**If cockpit visibility is dangerously restricted**

8. Canopy -- Jettison.

## PC HYDRAULIC FAILURE -- TOTAL

Total PC hydraulic failure is recognized by the cautions PC-1, PC-2, and FCS 2ND. Reduce hydraulic pressure demands by flying at airspeeds to minimize high pressure loads on the flight controls (approximately 220-300 KCAS).

1. Hydraulic Pressure -- Check.
Check hydraulic pressure on the standby flight display 2 to verify total PC failure.

2. Airspeed -- 220-300 KCAS.

3. Flight Controls -- Minimize inputs.

4. Controllability -- Check.
See CONTROLLABILITY CHECK.

5. Land as soon as possible.
Make straight in approach. Anticipate slow landing gear extension.

## PC AND UTILITY HYDRAULIC FAILURE

The loss of both PC and one utility hydraulic pump, or both utility and one PC pump, reduces flight control effectiveness to approximately half.

Loss of one PC and both utility hydraulic pumps automatically activates the emergency power unit and restores some flight control authority to the hydraulic system. With the loss of one utility and both PC pumps, select emergency power unit on. See EMERGENCY POWER OPERATION. See LANDING WITH EPU ACTIVATED.

1. EPU -- ON.

**If both PC hydraulic systems and one utility hydraulic pump failed**

2. Airspeed -- 170-180 KCAS.

3. Landing Gear Emergency Extension Switch DOWN.

#### NOTE

Lower the landing gear with emergency accumulator pressure to reduce hydraulic system demand. With both PC hydraulic systems failed, utility hydraulic system pressure could be insufficient to power the flight controls and utility system.

4. Land as soon as possible.
Make straight in approach.

## LOSS OF CANOPY

1. Airspeed -- Less than 200 KCAS.

2. Altitude -- Maintain less than 10,000 feet MSL.

3. Ejection Seat -- Lower.

4. Land as soon as practical.

## EJECTION GENERAL

Ejection can be accomplished from ground level to 50,000 feet MSL, between 0 and 600 knots with wings level and no sink rate. Appreciable forces are exerted on the body above 450 knots. Above 600 knots, ejection is hazardous due to excessive dynamic pressures. Altitude, airspeed, pitch and dive angles, sink rate, G loads, human reaction times, etc. can reduce chances for survival and could be cumulative. In most situations, ejection at higher altitudes (approximately 10,000 ft AGL) at reduced airspeeds allows more time to overcome ejection difficulties. In controlled level flight, eject above 2000 feet AGL if possible. If out-of-control, eject above 10,000 feet AGL if possible. Do not delay ejection below 2000 feet AGL for any reason.

WARNING

- To eliminate the risk of burn injury during ejection requires wearing flight gloves and flying with sleeves rolled down.

- Airblast injuries increase as ejection speeds increase. Reduce airspeed as much as practical before ejecting.

- Do not delay ejection for any reason that commits you to an unsafe ejection.

## HIGH ALTITUDE EJECTION

In an ejection at higher altitude, considerable time elapses during drogue chute descent before seat separation and parachute deployment. Therefore, manual seat separation should be made only if automatic seat separation failure is confirmed.

## LOW ALTITUDE EJECTION

Below 2,000 feet AGL, chances for a successful ejection are improved by leveling the wings and zooming the aircraft. Exchange excess airspeed for altitude and gain an upward vector for ejection. Do not exceed 20 degrees nose up. Eject before climb rate reaches zero. A more vertical seat trajectory provides additional time for automatic seat separation and parachute deployment.

## AUTOMATIC EJECTION SEQUENCING

There are three ejection modes, based on airspeed and altitude (figure 3-5). Mode 1 provides immediate parachute deployment after clearing the aircraft. Mode 2 deploys the drogue chute first to slow the seat, then deploys the parachute. Mode 3 is a high altitude/high airspeed mode. In mode 3 the drogue chute remains attached to the seat until mode 2 (or mode 1) parameters are met, then resumes the recovery sequence.

# BEFORE EJECTION

Proper body position minimizes ejection injury. Bank and dive angles increase the minimum ejection altitude (figure 3-6).

**If time and conditions permit**

1. Slow below 200 KCAS.

WARNING

If unable to slow below 200 KCAS, jettison canopy with canopy jettison handle prior to ejection to eliminate possibility of canopy collision with ejection seat or main parachute.

2. Tighten lap belt.

3. Fully insert oxygen mask bayonets.

4. Lower helmet visor.

5. Tighten chin strap.

6. Stow loose equipment.

7. Select IFF EMER.

8. Lock shoulder harness.

9. Notify controlling agency of location and intentions.

# CANOPY SEPERATION FAILURE

If the canopy fails to jettison after pulling the ejection handle, pull the other handle. If the canopy fails to jettison, maintain one hand on the ejection handle and attempt to jettison/open the canopy with other hand.

1. Canopy Jettison Handle -- Press unlock button and pull.

**If canopy fails to jettison**

2. Canopy Switch -- Open.

# EJECTION

Ejection should be accomplished at the lowest practical airspeed.

WARNING

- Sled test results indicated a potential for collisions between the canopy and the man/seat or recovery parachute exsists at airspeeds between 225-400 KCAS. If necessary to eject in controlled flight between 225-400 KCAS, jettisoning the canopy, wait 5 seconds, and then eject to eliminate hazard.

- Do not pull restraint emergency release handle prior to ejection. Activation of the handle disconnects lap belt and inertial reel harness. Safe ejection is impossible.

## EJECTION MODES

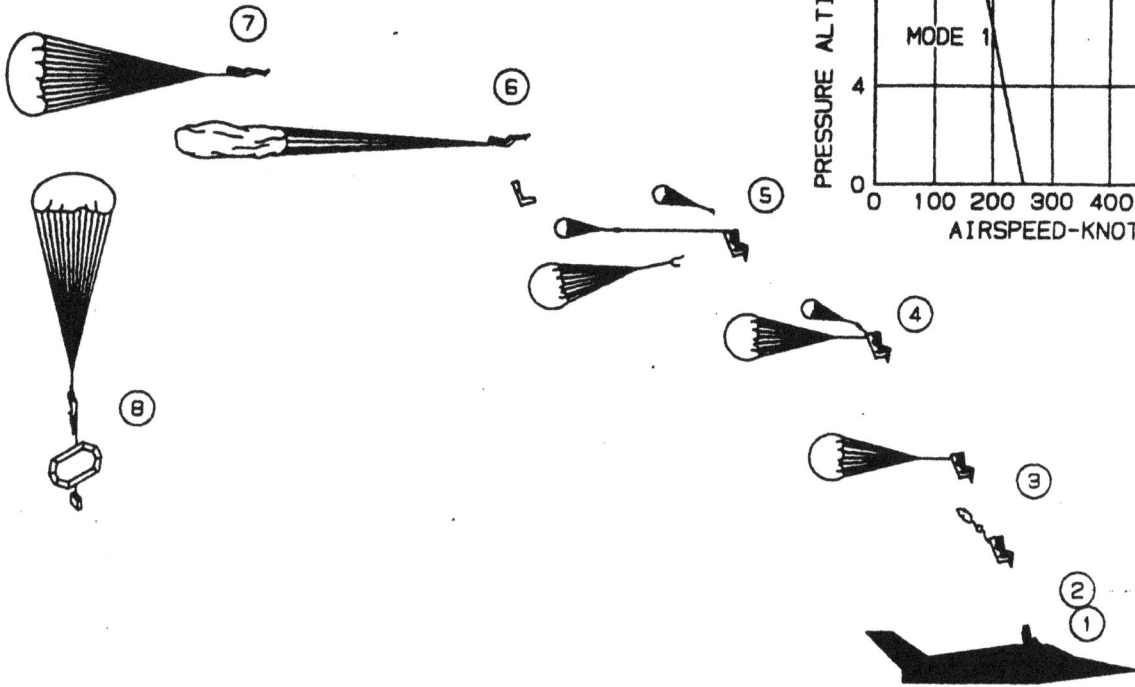

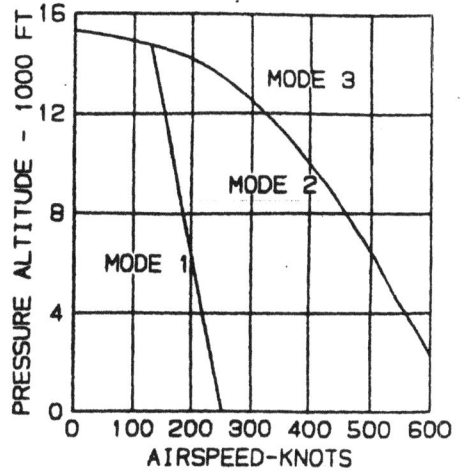

| TYPICAL EVENT TIMING | TIME (SECONDS) | | |
|---|---|---|---|
| | MODE 1 | MODE 2 | MODE 3 |
| 1. ROCKET CATAPULT FIRES | 0.0 | 0.0 | 0.0 |
| 2. VERNIER ROCKET FIRES | 0.18 | 0.18 | 0.18 |
| 3. DROGUE DEPLOYS AND INFLATES | NA | 0.17 | 0.17 |
| 4. PARACHUTE DEPLOYS | 0.20 | 0.97 | * |
| 5. DROGUE RELEASES FROM SEAT | NA | 1.32 | * |
| 6. SEAT RELEASES FROM PILOT | 0.45 | 1.42 | * |
| 7. PARACHUTE INFLATES | 1.80 | 2.8 | * |
| 8. SURVIVAL EQUIPMENT DEPLOYS | 5.8 | 6.3 | * |

\* PARACHUTE DEPLOYMENT, MAN SEAT SEPARATION AND FOL-
LOWING EVENTS ARE DELAYED UNTIL MODE 2 (OR MODE 1)
PARAMETERS ARE MET

C23-008

**Figure 3-5**

## EJECTION MINIMUM PARAMETERS

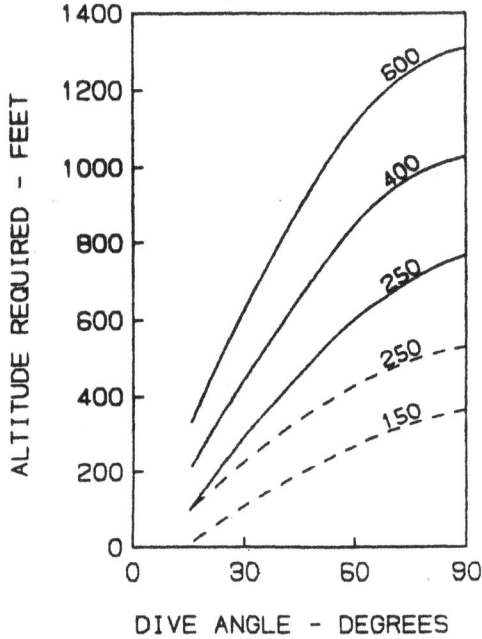

ALTITUDE VS DIVE ANGLE

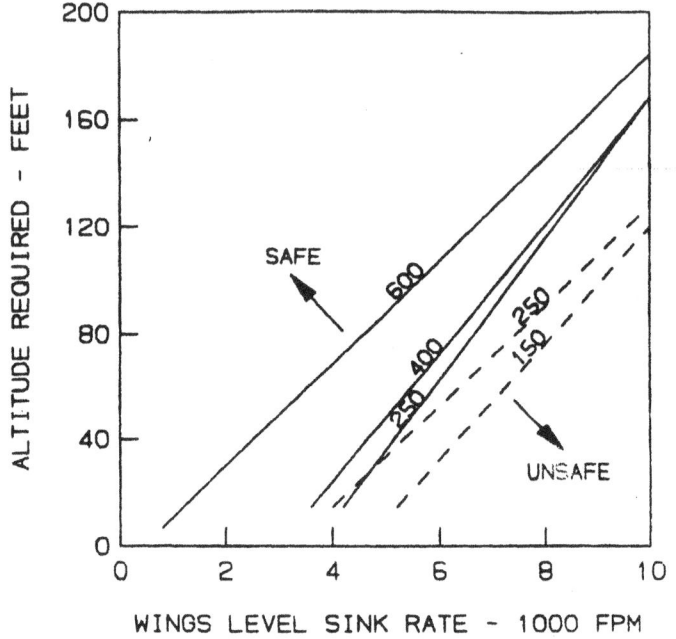

ALTITUDE VS SINK RATE

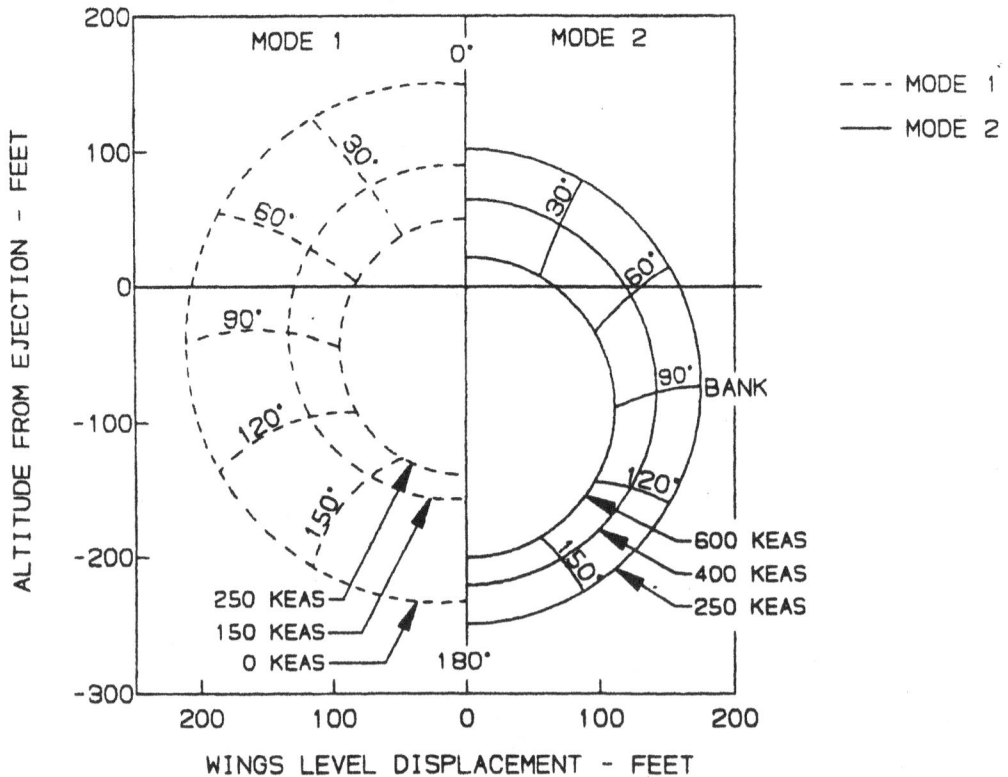

ALTITUDE VS BANK ANGLE

C23-008

Figure 3-6

1. Ejection Handle(s) -- Pull.

## EJECTION SEAT FAILURE

### WARNING

The following procedures are presented as a last resort action for the pilot to depart the aircraft and no in-depth studies have been made to confirm whether or not they will be successful. Pilot safety. may be jeopardized using these procedures; however this may be preferable to remaining with the aircraft.

If canopy separates and ejection seat does not fire.

1. Airspeed -- 200-250 KCAS.

2. Survival Kit Straps -- Release.

3. Trim -- Full Down
   Apply full down trim while hold aircraft in level flight.

### NOTE

Do not delay between step 4 and step 5.

4. Stick -- Release.

5. Restraint Emergency Release -- Actuate while pushing up on left Pitot probe support.

### MANUAL SEAT SEPARATION

Manual release is difficult and should be considered only as a last resort. Push up on the left Pitot support and simultaneously pull the restraint emergency release handle. Jerk on the parachute risers, or use any other method to force the personnel parachute container from the seat.

### WARNING

- Do not grasp the Pitot sensing inlet housing. Severe arm/hand injury could occur if parachute mortar fires.

- Do not attempt to open the lap belt. If the lap belt is opened, the seat partially falls away, but parachute risers and shoulder harness could remain attached and make successful parachute deployment impossible.

## EMERGENCY PROCEDURES
# AIR REFUELING

## BREAKAWAY PROCEDURES

Relative position of the tanker and receiver aircraft must be closely monitored by all crew members during all phases of refueling. If either a tanker or receiver crew member determines an abnormal condition requires an immediate separation of the two aircraft, the crew member shall transmit the breakaway call on air refueling frequency. These conditions include, but are not limited to, excessive rate of closure, overrun, and engine failure. The aircraft need not be in the contact position to call a breakaway.

Limit the use of the word breakaway to critical situations. Casual use of the word breakaway during refueling operations could cause aircrews to initiate emergency separation procedures.

For all breakaways, transmit the tanker call sign and the word breakaway three times (Example:*CHEVY 20, BREAKAWAY, BREAKAWAY, BREAKAWAY*).

**The tanker shall:**
- Increase power, if available, to obtain forward separation. When clear of the receiver, climb straight ahead with wings level.
- Turn beacon lights to BOTH ON and position lights to BRIGHT.

**The receiver shall:**
- Disconnect. After disconnect, retard throttles and establish a definite rate of descent to 1000 feet below base air refueling altitude.
- Drop aft of tanker until entire tanker is in sight and monitor flight instruments.
- If overrun of the tanker occurs, do not turn until positive separation is attained.

| WARNING |
|---|

If the receiver loses sight of the tanker at any time after overrun, the receiver shall establish a positive rate of descent to 1000 feet below base air refueling altitude.

## SYSTEMS MALFUNCTIONS

Do not attempt air refueling if systems malfunctions jeopardize safety except for a fuel emergency or command direction.

### FAILURE TO DISCONNECT

If a disconnect does not occur, select SLIPWAY ORIDE on the air refueling switch and push the air refueling disconnect switch. If a disconnect still does not occur, use a brute force disconnect.

### BRUTE FORCE DISCONNECT

A controlled brute force disconnect is gradual movement to the aft boom limit ending with a smooth tension pull out. A controlled brute force disconnect is a last resort.

Maintain refueling position and establish a gradual movement aft until the boom is at maximum extension. After separation, request the boom operator visually check for receptacle damage. Air refueling may be continued with other receivers if the boom shows no visual damage and a boom stability check is normal.

If an uncontrolled brute force disconnect occurs, air refueling shall be terminated except during fuel emergencies.

### OVERRIDE OPERATION

If inadvertent disconnects occur due to a tanker malfunctioning signal amplifier or receiver system malfunction, refueling can be completed by setting the air refueling switch to SLIPWAY ORIDE. With the switch in the override position, only the receiver aircraft can initiate an electrical disconnect. The tanker aircraft is not able to initiate a disconnect. Disconnects can still occur in the override position when the nozzle fuel pressure is over 82 psi, or a mechanical disconnect caused by exceeding hydraulic latch actuator relief setting.

| CAUTION |
|---|

During operation in the air refueling override mode, tanker-generated automatic disconnect is not available. Close coordination between the receiver pilot and the boom operator is required for disconnect to prevent boom and receptacle damage.

### FAILURE TO LATCH

When all other means for fuel transfer fail, and a fuel shortage emergency aboard the receiver aircraft exists, the boom operator can maintain boom contact with slight pressure on the receptacle.

## CAUTION

Prior to emergency transfer of fuel, the boom operator shall thoroughly coordinate the procedures to be used. After the required briefing, the receiver air refueling system is put into override. The boom operator inserts the boom nozzle fully into the receptacle and applies adequate fuel pressure to maintain boom/receptacle seal.

## TANKER ENGINE FAILURE

In the event tanker engine failure is experienced during air refueling, the refueling shall be terminated immediately. In fuel emergencies, every effort shall be made to onload fuel, utilizing a descending flight path if required.

## ABORT PROCEDURES

A tanker aborting refueling informs the receiver, clear the formation, and take action as required.

A receiver aborting during the refueling operation informs the tanker of the abort, clears the formation. and takes action as required.

# EMERGENCY PROCEDURES
# LANDING

## CONTROLLABILITY CHECK

Check controllability if structural damage is suspected or failures adversely affect aircraft handling characteristics. Adjust CG as necessary. Slow to gear lowering speed, lower landing gear, and reduce airspeed in 10 knot increments to an acceptable control speed or 10 degrees AOA.

1. Altitude -- Above 5000 feet AGL.
   Proceed to safe altitude and area. Maintain above minimum safe ejection altitude.

2. Landing Gear -- Down.
   If the structure is damaged, maintain landing configuration.

3. Airspeed -- Reduce.
   Slow gradually to an acceptable speed, but not less than approximately 140 KCAS, approximately 8-12 degrees AOA, or until an undesirable aircraft control problem is encountered.

4. Landing Approach -- Straight in.
   Fly final approach no slower than acceptable airspeed.

5. Touchdown -- Above minimum control airspeed.

## SINGLE ENGINE LANDING

A single engine provides sufficient power for a missed approach, if necessary. Consistent with safety, attempt to maintain failed engine at idle to avoid loss of hydraulic and electric redundancy. Plan approach to avoid high thrust settings.

| WARNING |

Use of afterburner aggravates directional control problems and increases minimum control airspeed.

1. Gross Weight -- Reduce.

2. Nonessential Electrical Equipment -- Off.

3. Fly straight in approach.

## NO FLAP LANDING

If a landing is to be made with the wing flaps retract-retracted, landing distance is increased due to higher touchdown speed and less effective aerodynamic braking.

1. Fly straight in approach.

2. Airspeed/AOA -- Maintain TBD knots/TBD degrees.
   Maintain approximately 10 KCAS above normal final approach speed.

## PRECAUTIONARY LANDING PATTERN

The type of pattern flown in an emergency depends on type of malfunction, weather conditions, suitability and proximity of a landing field, and fuel remaining. The landing pattern is a straight-in approach, a higher than normal straight-in approach, or flameout pattern.

### STRAIGHT-IN APPROACH

A straight-in approach is recommended for emergencies that require minimum maneuvering (hydraulic, flight control, or electrical problems) or situations that require high thrust levels. A controllability check should be accomplished prior to the approach if control difficulties are anticipated. Airspeed and altitude should be maintained as long as possible to insure safe ejection parameters during the approach.

## FLAMEOUT LANDING

The decision to eject or make a flameout landing rests with the pilot. Other emergency conditions and proficiency in preforming flameout approach and landings should be considered.

Do not delay an ejection in an attempt to salvage a questionable approach.

With the EPU operating, the standby attitude indicator is the only attitude reference. The right MPCD, flight test equipment, and communications should be operative.

If the EPU is not operating, immediately check the the EPU control switch is ON. If the EPU is not operative, maintain sufficient airspeed to provide hydraulic pressures. The VMS battery power is approximately 2 minutes without the EPU running. If the EPU cannot be activated, a safe landing is improbable and a controlled ejection is recommended.

Flaps must be manually selected and the landing gear extended using the emergency system. Emergency gear extension requires approximately 25-35 seconds. Emergency brakes and nosewheel steering should be selected prior to touchdown.

To start a flameout approach, turn immediately toward the landing runway and establish a best range airspeed (or AOA). The EPU must be on and should provide approximately nine minutes of operating time.

The basic flameout approach is an overhead pattern or an alternate straight-in approach when an overhead pattern can not be attained. The overhead pattern can be entered at any point along the path, if the prescribed altitude can be attained. The initial aim point for touchdown is approximately 1/3 of the way from the approach end of the runway.

### OVERHEAD APPROACH

Plan the approach to arrive over the runway (high key) at 10,000-12,000 feet AGL. The high key can be entered from any direction. The minimum high key altitude of 10,000 feet AGL is based on flying a 360 degree pattern, landing gear down, and pattern airspeed of 220 KCAS. If aircraft is above 12,000 feet at high key, fly an altitude dissipating maneuver. Avoid rapid control inputs that will deplete EPU fuel and could exceed the capability of the emergency hydraulic pumps. If EPU fuel quantity is below 30 percent at high key, a flameout landing should not be attempted since adequate hydraulic pressure may not be available through landing.

Plan the approach to arrive at low key (abeam) the runway at 8,000 feet AGL. Fly a ground track from low key to arrive on a 1/2 mile final. Lowering the landing gear can be delayed if required to conserve energy.

### STRAIGHT-IN APPROACH

If the overhead pattern positions cannot be made, a straight-in approach can be flown. Maintain a clean configuration glide until the runway touchdown point appears to be 12-15 degrees below the horizon. Lower the landing gear, set the flap configuration, and maintain 220 KCAS approach airspeed.

### LANDING PHASE

Landing gear should be lowered no later than 4,000 feet AGL to allow adequate extension time. Main-tain 220 KCAS until final, maintain minimum of 200 KCAS on final approach. Once landing is assured, lower nose, shift aimpoint to beginning of overrun, and allow airspeed to increase. Start the flare to touchdown with a normal landing sink rate.

> **WARNING**
>
> Do not delay ejection to below 2,000 feet MSL in an attempt to salvage a questionable landing approach.

1. Emergency Oxygen Bottle Handle -- Activate.

2. Airspeed -- 220 KCAS.

3. EPU Switch -- ON.

4. Landing Gear Emergency Extension Switch -- DOWN.

> **WARNING**
>
> Do not delay lowering landing gear below 4,000 feet MSL.

5. Flaps -- As required.

6. Emergency Brake/Steering Switch -- ON.

7. Airspeed -- 200 KCAS (optimum).

**After stop**

8. EPU Switch -- OFF.

## LANDING WITH UTILITY HYDRAULIC FAILURE

Utility hydraulic system failure can be recognized by both the cautions UTIL 1 and UTIL 2, or the single caution UTIL B. Emergency accumulator pressure is required to lower landing gear and to operate brakes and nosewheel steering.

Consider normal extension of the landing gear when either utility hydraulic system is lost or failing.

Antiskid protection is not available with use of the emergency braking system. If hot brakes suspected, refer to HOT BRAKES.

**Conditions permitting**

1. Gear Handle -- DN.

## LANDING GEAR EMERGENCY EXTENSION

1. Airspeed -- 170-180 KCAS.

2. Gear Handle -- DN.
   If remaining hydraulic pressure is sufficient to extend and lock the gear down normally, landing gear emergency extension may not be required.

3. Landing Gear Emergency Extend Switch -- DOWN.

### NOTE

If the emergency landing gear system fails to fully extend the gear, reduce hydraulic demand and allow emergency accumulator to recharge. When the the caution APU ACCUM rescinds, attempt another landing gear emergency extension.

### After touchdown

4. Brakes -- Check.

5. Emergency Brake Steering Switch -- BRAKE. Limit the number of nosewheel steering inputs. Antiskid is inoperative.

## ABNORMAL GEAR CONFIGURATION

If all landing gear cannot be extended with the normal system, attempt to extend the gear using the landing gear emergency extension. If unable to get all gear down and locked, use Northrop engineering guide for landing with abnormal gear configuration, ABNORMAL GEAR CONFIGURATION GUIDE, figure 3-7. If both main gear cannot be retracted or extended, consider ejection.

## LANDING WITH TIRE FAILURE

If landing with a flat nose or main gear tire, fly a normal pattern. Plan to touch down in the center of the runway.

### Flat MLG tire

After touchdown, lower nosewheel to runway, engage nosewheel steering, and use a combination of flight control, nosewheel steering, and brakes to maintain directional control.

1. Antiskid Switch -- OFF.

### Flat nose wheel tire

After touchdown, hold nose wheel off runway to minimize FOD to the engine. Lower nose gently to the runway while sufficient elevator control authority is available. After nose touches down, hold back stick pressure, and use moderate braking to stop the aircraft. Nosewheel steering is automatically engaged with weight on wheels and requires holding the paddle switch to disengage nosewheel steering.

## LANDING AFTER EPU ACTIVATED

If the EPU switch is left in the AUTO position, the EPU system automatically turns off five seconds after landing. If the EPU is required for landing, ensure EPU is ON.

Notify ground personnel and request biological/environmental services support. Select 100% oxygen. After landing, turn off ECS and activate emergency oxygen bottle. Park in an isolated, controlled area. Treat all leaks as hydrazine leaks. Turn EPU selector off and request EPU ground safety switch be turned off.

1. Oxygen Power Switch -- 100%.

2. EPU -- ON.

### After landing

3. Emergency Oxygen Bottle Handle -- Activate. Provides 5-15 minutes of oxygen.

4. Oxygen Power Switch -- POWER OFF.

5. Oxygen Supply Switch -- OFF.

6. Cabin Pressure Switch -- RAM/DUMP.

7. EPU Selector -- Off.

8. Aircraft -- Park into the wind. Park aircraft in an isolated area.

# LANDING WITH ABNORMAL GEAR CONFIGURATION
## BEFORE LANDING CONSIDERATIONS

1. JETTISON ARMAMENT
2. REDUCE WEIGHT TO 5000 LB FUEL MAX          4. FLY FLAT APPROACH

| | LANDING NOT RECOMMENDED |
|---|---|
| ONE MAIN-NO NOSE | ● RETRACT GEAR AND REFER TO ALL GEAR UP<br>IF GEAR WILL NOT RETRACT-<br>● RECOMMEND EJECT |
| ONE MAIN-ONE NOSE DOWN | ● RETRACT GEAR AND REFER TO ALL GEAR UP<br>IF GEAR WILL NOT RETRACT-<br>● RECOMMEND EJECT |
| | LAKEBED LANDING RECOMMENDED |
| TWO MAIN-NO NOSE | ● ACTIVATE EPU ON FINAL APPROACH<br>● BE PREPARED TO COUNTER NOSE DIP<br>● 110 KNOTS LOWER NLG SMOOTHLY TO LAKEBED<br>● SHUT DOWN BOTH ENGINES AFTER NOSE TOUCHDOWN<br>● SELECT EMERGENCY BRAKES AFTER ENGINE SHUTDOWN |
| ALL GEAR UP | ● 180 KNOTS TOUCHDOWN SPEED |
| NO MAIN-NOSE DOWN | ● BE PREPARED TO COUNTER WING DIP<br>● DO NOT SHUT DOWN ENGINES UNTIL STOPPED |
| | RUNWAY LANDING RECOMMENDED |
| NOSE WHEEL MISSING | ● ACTIVATE EPU ON FINAL APPROACH<br>● MAKE NORMAL APPROACH AND 2 PT TOUCHDOWN<br>● 110 KNOTS LOWER NLG SMOOTHLY TO RUNWAY<br>● SHUT DOWN BOTH ENGINES AFTER NLG TOUCHDOWN<br>● SELECT EMERGENCY BRAKES AFTER ENGINE SHUTDOWN |
| NO MAIN WHEEL-BRAKE INTACT | ● ANTI-SKID - OFF<br>● BE PREPARED TO COUNTER WING DIP<br>● LAND ON SIDE OF RUNWAY TOWARD GOOD GEAR<br>● USE NOSEWHEEL STEERING AND GOOD BRAKE TO MAINTAIN DIRECTIONAL CONTROL |

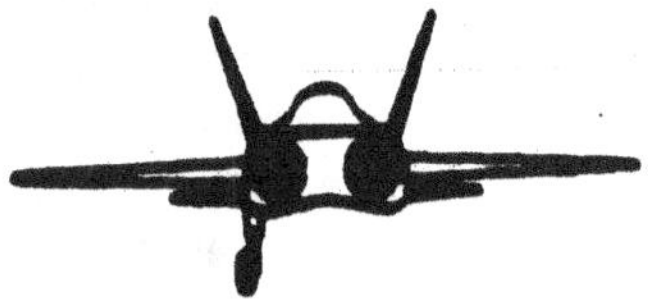

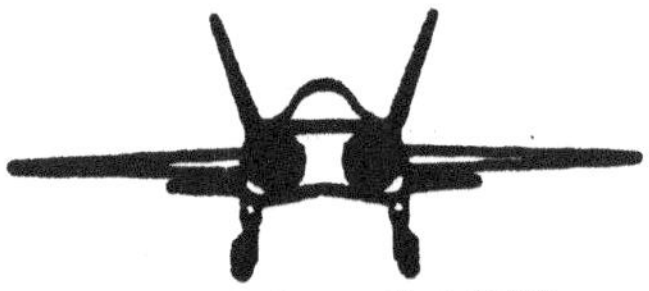

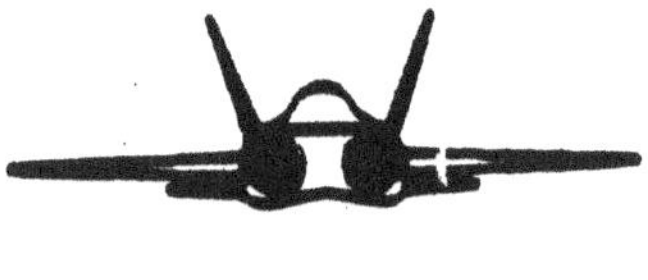

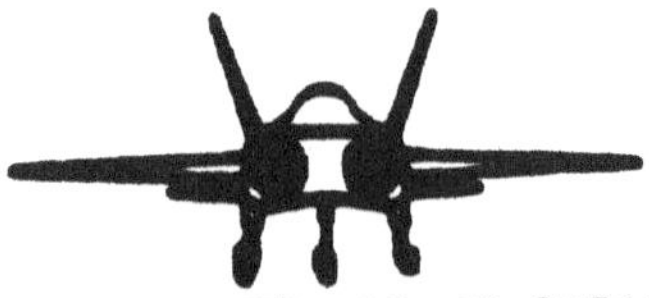

Figure 3-7

## SECTION V

## OPERATING LIMITATIONS

E23-SECV

## Supplement Flight and Operating Limitations Manual

Operating limitations are provided by Supplement Flight and Operating Limitations Manual, published by Northrop System Safety (Orgn 3891/89).

# SECTION VI
## FLIGHT CHARACTERISTICS

F23-SECVI

# FLIGHT CHARACTERISTICS

Flight charaterisics to be included when available.

# SECTION VII

## ADVERSE WEATHER OPERATIONS

029-SECVII

# TABLE OF CONTENTS

## COLD WEATHER OPERATION

The ground crews accomplish aircraft preparation for flight. The aircrew should recheck items that are peculiar to cold weather.

### BEFORE ENTERING AIRCRAFT

Remove protective covers.

Check aircraft for ice, snow, or slush accumulations. Have deiced, as necessary.

Thrust required to start taxi could be substantially higher than normal. Ensure area behind aircraft is clear.

### BEFORE ENGINE START

Use ground powered heater to preheat cockpit.

### ENGINE START

Oil pressure could rise rapidly above normal operating range if the engines have been cold soaked. Do not operate the engine above idle until oil pressure is within normal operating range.

White smoke from unburned fuel vapor could appear coming from the engine exhausts during start if the aircraft has been cold soaked for an extended period of time.

Turn on defog to ensure the windscreen is clear when ready to taxi.

### TAKEOFF

Ice, snow and frost must be removed from aircraft surfaces prior to takeoff. Stall speeds, flight characteristics, and aircraft performance are affected adversely and unpredictably. Request deicing if necessary. If deicing fluid is sprayed directly into the engine inlets, turn off the cabin conditioning system and turn on windshield anti-ice for five minutes to purge the bleed air system of deicing fluid fumes.

Ice and light accumulations of powdery snow have little effect on takeoff distance, but can adversely affect directional control and takeoff.

If runway is covered with standing water or slush, leave landing gear down for approximately 30 seconds after takeoff to allow moisture to blow off the landing gear and gear doors.

### ENROUTE

Surface icing inflight reduces range and climb performance. Increased power is required to maintain desired airspeed.

The pitot static heaters are not operational with a complete ac power failure. Do not fly through known or suspected icing conditions with complete ac power failure. Iced-over air data static ports create erroneous air data inputs to the flight control computers.

## APPROACH/LANDING

Turn on defog 15 minutes prior to descent.

On wet or icy runways, optimum anti-skid braking is obtained with smoothly applied and steady brake pressure. Low braking coefficients at high speed on slippery runways cause a gradual buildup of deceleration rate. The anti-skid system goes through several skid cycles to establish the correct amount of brake pressure. The cycling is more noticeable at slower speeds. The total stopping distance with anti-skid is shorter than without anti-skid.

## SHUTDOWN

Allow engines to idle for 5 minutes. Allowing the engines to stabilize at idle temperature prevents cold shock to the engines after shutdown.

# HOT WEATHER AND DESERT OPERATION

Hot weather operations include dry or tropical climates.

Cool cockpit.

If sand or dirt is blowing, keep canopy closed.

Be sure engine inlets, air data ports, and landing gear struts are free of sand and dirt.

Inspect struts and hydraulic actuators for leaks from dried hydraulic seals.

Ensure area behind aircraft is clear so blowing sand or dirt does not affect other aircraft and equipment after engine start.

## BEFORE ENGINE START

Clean instrument panels of sand, dirt, or moisture.

## ENGINE START

Do not delay takeoff in blowing dust or dirt. Delay with engines running could increase engine wear substantially.

Increase ECS temperature until cockpit fog is eliminated.

## TAXI

Aircraft gross weight, taxi speed, and taxi distance influence brake temperature and rate of heat rise. Intermittent but positive braking results in cooler brakes and less wear. Use the largest turning radius conditions permit.

## POSTFLIGHT

Release parking brake after wheel chocks are in place.

Park aircraft crosswind and install protective covers to preclude blowing sand and dirt from entering engine inlets or exhausts.

# APPENDIX

# PERFORMANCE DATA

## PERFORMANCE CONVERSION CHARTS

Standard conversion charts are provided for:

- Temperature Conversion, figure P-1.
- Cross Wind Chart, figure P-2.
- Standard Atmosphere, figure P-3.
- Airspeed VS Mach, figure P-4.

## ATF PERFORMANCE CHARTS

Performance charts for the ATF are provided under seperate cover by Northrop ATF Performance Engineering (Orgn 3732/89).

# TEMPERATURE CONVERSION

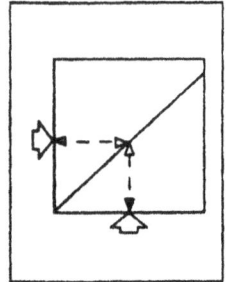

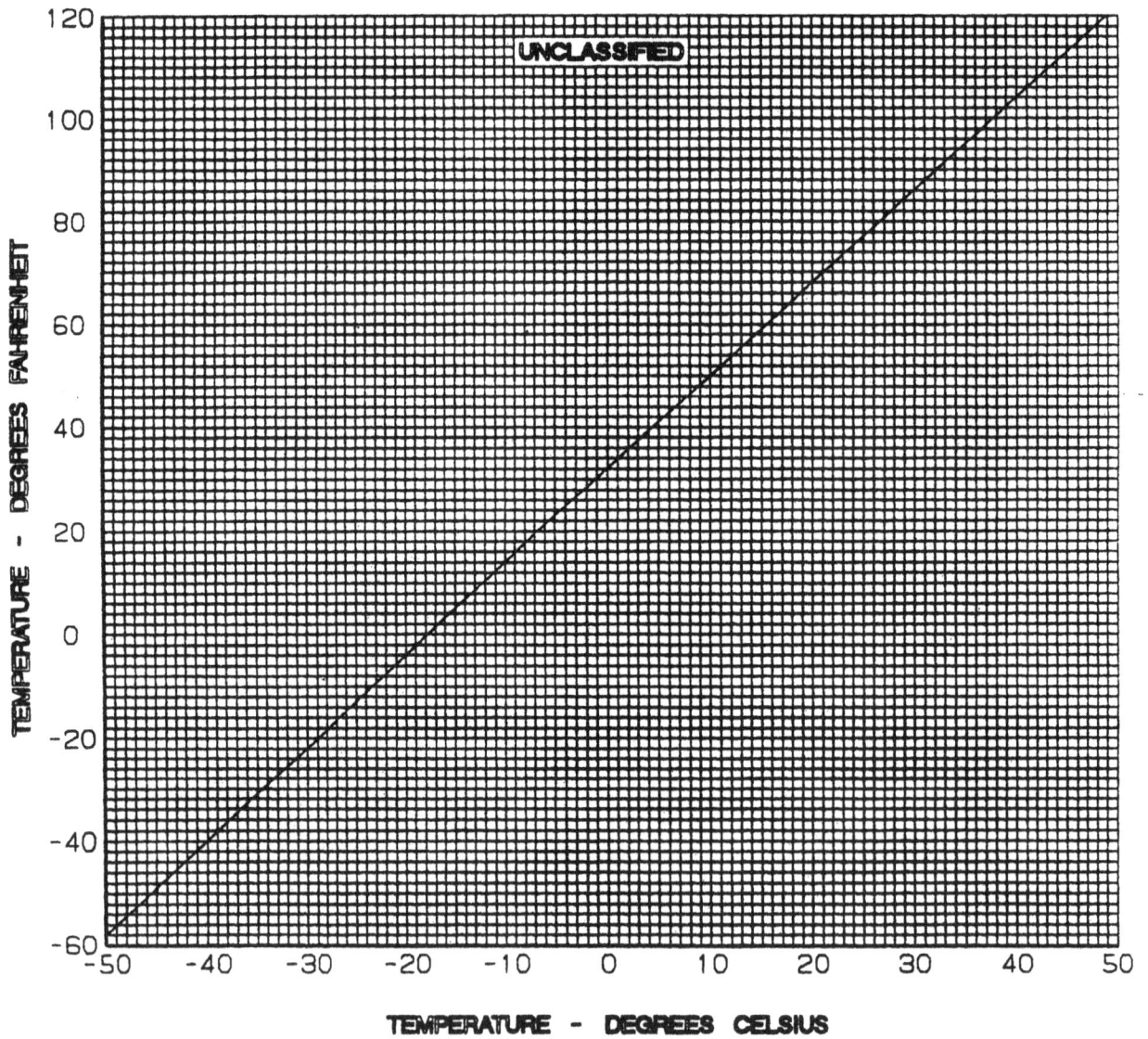

Figure P-1

P23-010

## CROSS WIND CHART.

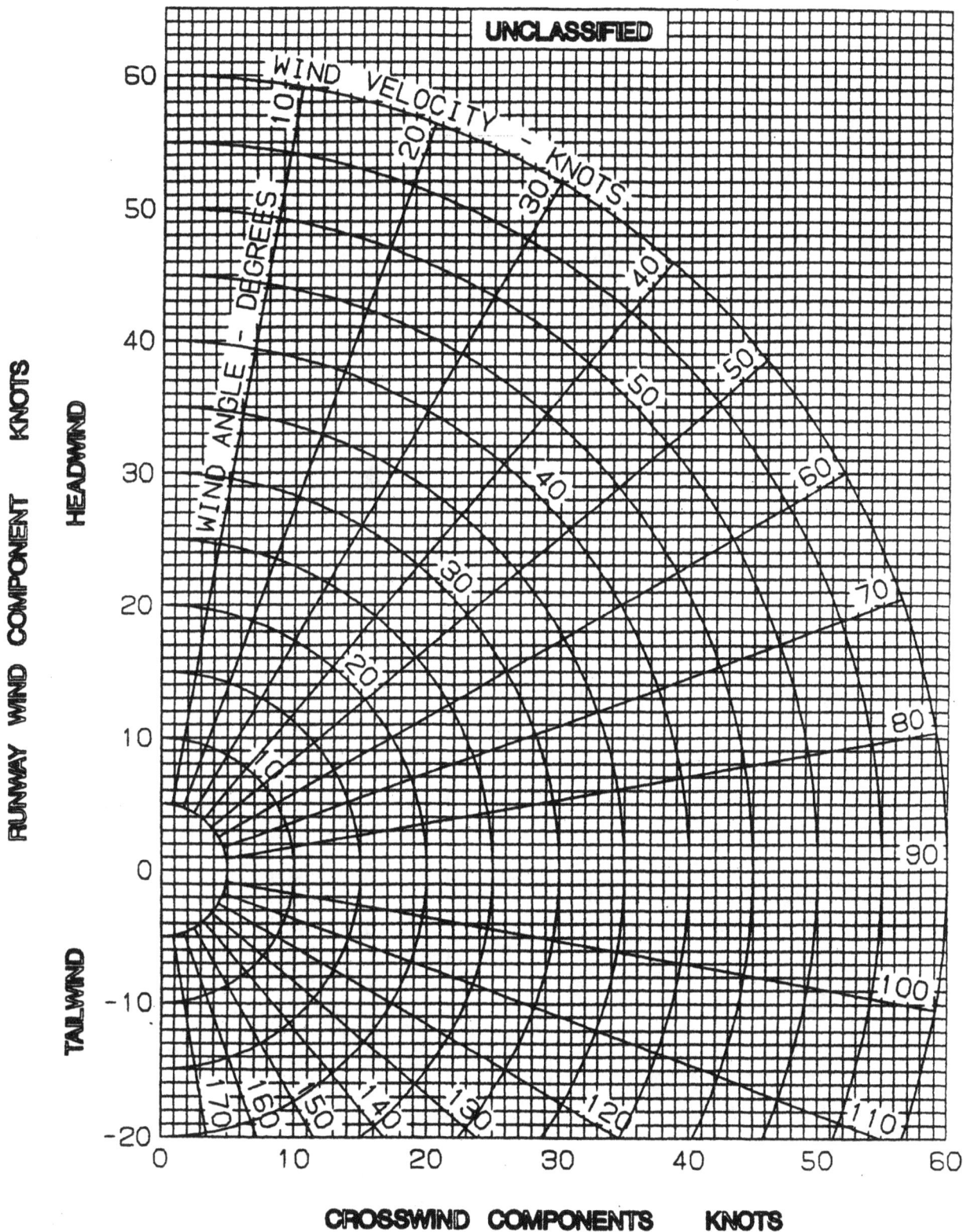

Figure P-2

**NORTHROP/MCAIR YF-23A**
**COMPETITION SENSITIVE**

## STANDARD ATMOSPHERE

STANDARD SEA LEVEL AIR:
T=59°F (15°C)
P=29.921 IN. OF HG

W = 0.076475 LB/CU FT  $\rho_o$ = 0.0023769 SLUGS/CU FT
1 IN. OF HG = 70.732 LB/SQ FT = 0.4912 LB/SQ FT
$a_o$=1116.44 FT/SEC = 661.47 KT

| ALTITUDE FEET | DENSITY RATIO $\rho/\rho_o=\sigma$ | $1/\sqrt{\sigma}$ | TEMPERATURE | | SPEED OF SOUND RATIO $a/a_o$ | PRESSURE | |
|---|---|---|---|---|---|---|---|
| | | | DEG F | DEG C | | IN. OF HG | RATIO $P/P_o=\delta$ |
| -2000 | 1.0598 | 0.8714 | 66.132 | 18.962 | 1.0064 | 32.15 | 1.0744 |
| -1000 | 1.0296 | 0.8855 | 62.566 | 16.981 | 1.0030 | 31.02 | 1.0367 |
| 0 | 1.0000 | 1.0000 | 59.000 | 15.000 | 1.0000 | 29.92 | 1.0000 |
| 1000 | 0.9711 | 1.0148 | 55.434 | 13.018 | 0.9966 | 28.86 | 0.9644 |
| 2000 | 0.9428 | 1.0299 | 51.868 | 11.038 | 0.9931 | 27.82 | 0.9298 |
| 3000 | 0.9151 | 1.0454 | 48.302 | 9.057 | 0.9896 | 26.82 | 0.8962 |
| 4000 | 0.8881 | 1.0611 | 44.735 | 7.075 | 0.9862 | 25.84 | 0.8637 |
| 5000 | 0.8617 | 1.0773 | 41.169 | 5.094 | 0.9827 | 24.90 | 0.8320 |
| 6000 | 0.8359 | 1.0938 | 37.603 | 3.113 | 0.9792 | 23.98 | 0.8014 |
| 7000 | 0.8106 | 1.1107 | 34.037 | 1.132 | 0.9756 | 23.09 | 0.7716 |
| 8000 | 0.7860 | 1.1279 | 30.471 | -0.849 | 0.9721 | 22.22 | 0.7428 |
| 9000 | 0.7620 | 1.1456 | 26.905 | -2.831 | 0.9686 | 21.39 | 0.7148 |
| 10,000 | 0.7385 | 1.1637 | 23.339 | -4.812 | 0.9650 | 20.58 | 0.6877 |
| 11,000 | 0.7156 | 1.1822 | 19.772 | -6.793 | 0.9614 | 19.79 | 0.6614 |
| 12,000 | 0.6932 | 1.2011 | 16.206 | -8.774 | 0.9579 | 19.03 | 0.6360 |
| 13,000 | 0.6713 | 1.2205 | 12.640 | -10.756 | 0.9543 | 18.29 | 0.6113 |
| 14,000 | 0.6500 | 1.2403 | 9.074 | -12.737 | 0.9507 | 17.58 | 0.5875 |
| 15,000 | 0.6292 | 1.2606 | 5.508 | -14.718 | 0.9470 | 16.88 | 0.5643 |
| 16,000 | 0.6090 | 1.2815 | 1.941 | -16.669 | 0.9434 | 16.22 | 0.5420 |
| 17,000 | 0.5892 | 1.3028 | -1.625 | -18.681 | 0.9397 | 15.57 | 0.5203 |
| 18,000 | 0.5699 | 1.3246 | -5.191 | -20.662 | 0.9361 | 14.94 | 0.4994 |
| 19,000 | 0.5511 | 1.3470 | -8.757 | -22.643 | 0.9324 | 14.34 | 0.4791 |
| 20,000 | 0.5328 | 1.3700 | -12.323 | -24.624 | 0.9287 | 13.75 | 0.4595 |
| 21,000 | 0.5150 | 1.3935 | -15.889 | -26.605 | 0.9250 | 13.18 | 0.4406 |
| 22,000 | 0.4976 | 1.4176 | -19.456 | -28.587 | 0.9213 | 12.64 | 0.4223 |
| 23,000 | 0.4807 | 1.4424 | -23.022 | -30.568 | 0.9175 | 12.11 | 0.4046 |
| 24,000 | 0.4642 | 1.4678 | -26.588 | -32.549 | 0.9138 | 11.60 | 0.3876 |
| 25,000 | 0.4481 | 1.4938 | -30.154 | -34.530 | 0.9100 | 11.10 | 0.3711 |
| 26,000 | 0.4325 | 1.5206 | -33.720 | -36.511 | 0.9062 | 10.63 | 0.3552 |
| 27,000 | 0.4173 | 1.5480 | -37.286 | -38.492 | 0.9024 | 10.17 | 0.3398 |
| 28,000 | 0.4025 | 1.5762 | -40.852 | -40.473 | 0.8986 | 9.725 | 0.3250 |
| 29,000 | 0.3881 | 1.6052 | -44.419 | -42.455 | 0.8948 | 9.297 | 0.3107 |
| 30,000 | 0.3741 | 1.6349 | -47.985 | -44.436 | 0.8909 | 8.885 | 0.2970 |
| 31,000 | 0.3605 | 1.6654 | -51.551 | -46.417 | 0.8871 | 8.498 | 0.2837 |
| 32,000 | 0.3473 | 1.6968 | -55.117 | -48.398 | 0.8832 | 8.106 | 0.2709 |
| 33,000 | 0.3345 | 1.7291 | -58.683 | -50.379 | 0.8793 | 7.737 | 0.2586 |
| 34,000 | 0.3220 | 1.7623 | -62.249 | -52.361 | 0.8754 | 7.382 | 0.2467 |
| 35,000 | 0.3099 | 1.7964 | -65.816 | -54.342 | 0.8714 | 7.041 | 0.2353 |
| 36,000 | 0.2981 | 1.8315 | -69.382 | -56.323 | 0.8675 | 6.712 | 0.2243 |
| 37,000 | 0.2844 | 1.8753 | -69.700 | -56.500 | 0.8671 | 6.397 | 0.2138 |
| 38,000 | 0.2710 | 1.9209 | -69.700 | -56.500 | 0.8671 | 6.097 | 0.2038 |
| 39,000 | 0.2583 | 1.9677 | -69.700 | -56.500 | 0.8671 | 5.811 | 0.1942 |
| 40,000 | 0.2462 | 2.0155 | -69.700 | -56.500 | 0.8671 | 5.538 | 0.1851 |
| 41,000 | 0.2346 | 2.0645 | -69.700 | -56.500 | 0.8671 | 5.278 | 0.1764 |
| 42,000 | 0.2236 | 2.1148 | -69.700 | -56.500 | 0.8671 | 5.030 | 0.1681 |
| 43,000 | 0.2131 | 2.1662 | -69.700 | -56.500 | 0.8671 | 4.794 | 0.1602 |
| 44,000 | 0.2031 | 2.2189 | -69.700 | -56.500 | 0.8671 | 4.569 | 0.1527 |
| 45,000 | 0.1936 | 2.2728 | -69.700 | -56.500 | 0.8671 | 4.355 | 0.1455 |
| 46,000 | 0.1845 | 2.3281 | -69.700 | -56.500 | 0.8671 | 4.151 | 0.1387 |
| 47,000 | 0.1758 | 2.3848 | -69.700 | -56.500 | 0.8671 | 3.956 | 0.1322 |
| 48,000 | 0.1676 | 2.4428 | -69.700 | -56.500 | 0.8671 | 3.770 | 0.1260 |
| 49,000 | 0.1597 | 2.5022 | -69.700 | -56.500 | 0.8671 | 3.593 | 0.1201 |
| 50,000 | 0.1522 | 2.5630 | -69.700 | -56.500 | 0.8671 | 3.425 | 0.1145 |
| 51,000 | 0.1451 | 2.6254 | -69.700 | -56.500 | 0.8671 | 3.264 | 0.1091 |
| 52,000 | 0.1383 | 2.6892 | -69.700 | -56.500 | 0.8671 | 3.111 | 0.1040 |
| 53,000 | 0.1318 | 2.7546 | -69.700 | -56.500 | 0.8671 | 2.965 | 0.09909 |
| 54,000 | 0.1256 | 2.8216 | -69.700 | -56.500 | 0.8671 | 2.826 | 0.09444 |
| 55,000 | 0.1197 | 2.8903 | -69.700 | -56.500 | 0.8671 | 2.693 | 0.09001 |
| 56,000 | 0.1141 | 2.9606 | -69.700 | -56.500 | 0.8671 | 2.567 | 0.08578 |
| 57,000 | 0.1087 | 3.0326 | -69.700 | -56.500 | 0.8671 | 2.446 | 0.08176 |
| 58,000 | 0.1036 | 3.1063 | -69.700 | -56.500 | 0.8671 | 2.331 | 0.07792 |
| 59,000 | 0.09877 | 3.1818 | -69.700 | -56.500 | 0.8671 | 2.222 | 0.07426 |
| 60,000 | 0.09414 | 3.2593 | -69.700 | -56.500 | 0.8671 | 2.118 | 0.07078 |
| 61,000 | 0.08972 | 3.3386 | -69.700 | -56.500 | 0.8671 | 2.018 | 0.06746 |
| 62,000 | 0.08551 | 3.4198 | -69.700 | -56.500 | 0.8671 | 1.924 | 0.06429 |
| 63,000 | 0.08150 | 3.5029 | -69.700 | -56.500 | 0.8671 | 1.833 | 0.06127 |
| 64,000 | 0.07767 | 3.5881 | -69.700 | -56.500 | 0.8671 | 1.747 | 0.05840 |
| 65,000 | 0.07403 | 3.6754 | -69.700 | -56.500 | 0.8671 | 1.665 | 0.05566 |

P23-012

**Figure P-3**

**NORTHROP/MCAIR YF-23A**
**COMPETITION SENSITIVE**

## AIRSPEED VS MACH

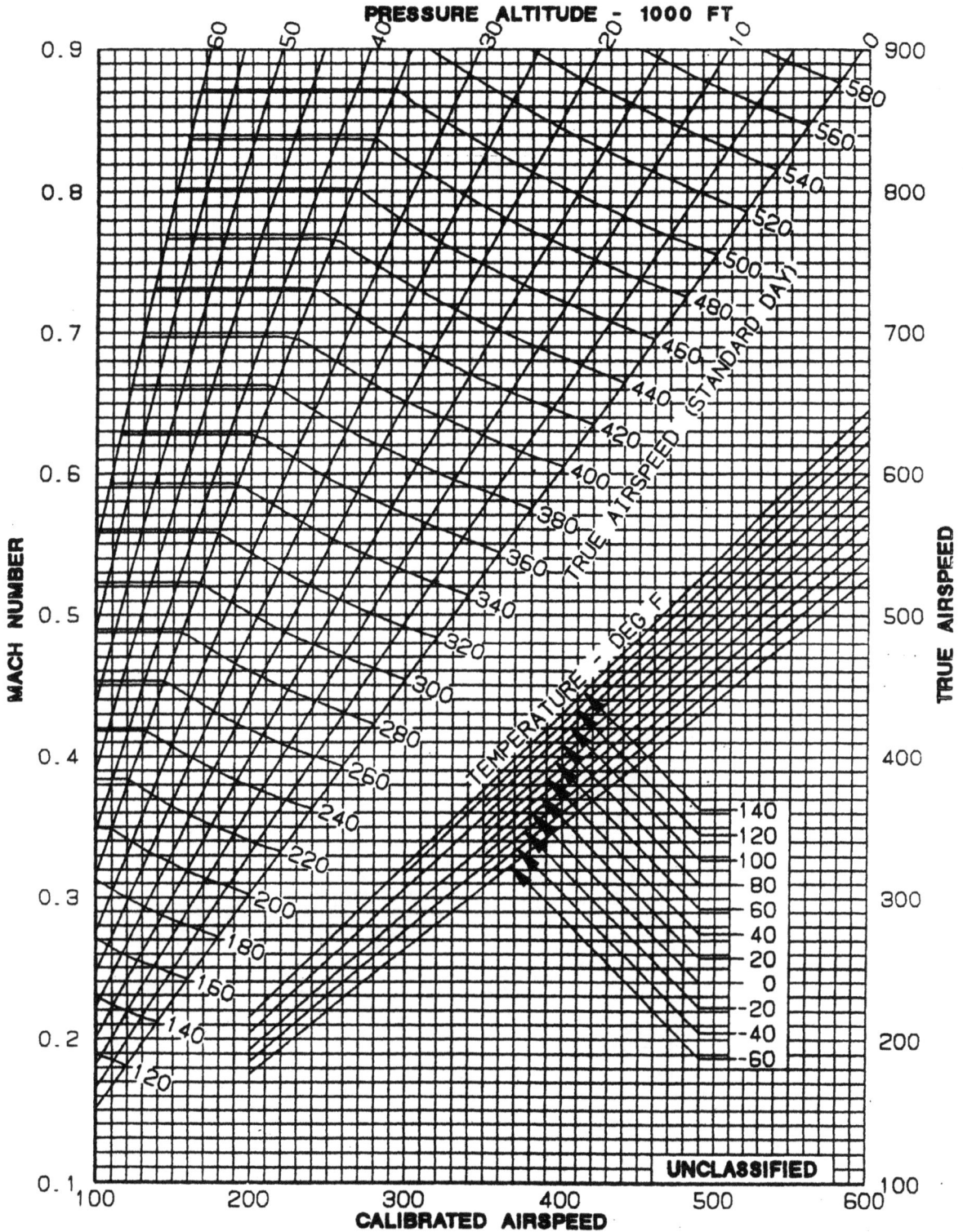

P23-013

*Figure P-4*

**NORTHROP/MCAIR YF-23A**
**COMPETITION SENSITIVE**
**UNCLASSIFIED**

A-5(A-6 blank)

# GLOSSARY

A29-GLOSS

# ACRONYMS AND ABBREVIATIONS

| | | | | |
|---|---|---|---|---|
| 1ST | First | | ATF | Advanced Tactical Fighter |
| 2DCD | Two-Dimensional, Convergent/Divergent | | ATL | Advanced Technology Launcher |
| 2ND | Second | | ATR | Automatic Trim Record |
| 3RD | Third | | ATS | Air Turbine Starter |
| | | | AUG | Augment |
| | **A** | | AUTO | Automatic |
| | | | AUX | Auxiliary Power |
| A | Automatic | | AV | Avionics |
| A/A | Air-to-Air | | | |
| A/D | Air Data | | | **B** |
| A/G | Air-to-Ground | | | |
| A/P | Auto Pilot | | B | Bingo Fuel |
| AAR | Air To Air Refueling | | BATT | Battery |
| ac | Alternating Current | | BB | Baseband |
| ACCEL | Accelerate | | BBW | Brake-By-Wire |
| ACCUM | Accumulator | | BIT | Built-In Test |
| ACES | Advanced Concept Ejection Seat | | BK | Brake |
| ACT | Actuator | | BLC | Boundary Layer Control |
| ADI | Attitude Director Indicator | | BLD | Bleed |
| AFXFR | Aft Tank Transfer | | BOT | Bottom |
| AGI | Armament Gas Ingestion | | BRG | Bearing |
| AIM | Air Intercept Missile | | BRT | Brightness |
| AIU | Avionics Interface Units | | BST | Boost |
| AJ | Antijam | | | |
| alpha | Angle-Of-Attack | | | **C** |
| ALT | Alternate | | | |
| | Altitude | | C | Calibrated Airspeed |
| AMAD | Airframe Mounted Accessory Drive | | | Centigrade Celsius |
| AMRAAM | Advanced Medium Range Air-To-Air | | | Cipher |
| | Missile | | C/M | Channel or Manual Frequency |
| ANT | Antenna | | CAS | Calibrated Airspeed |
| AOA | Angle Of Attack or ALPHA | | CC | Central Computer |
| AOS or | | | CDI | Course Deviation Indicator |
| AOSS | Angle of Sideslip or BETA | | CEPR | Core Engine Pressure Ratio |
| APU | Auxiliary Power Unit | | CG | Center of Gravity |
| ARMT | Armament | | CL | Closed |
| ARR | Air Refueling Receptacle | | | Clear |
| ASYM | Asymmetric | | CLG | Cooling |

| | | | | |
|---|---|---|---|---|
| CLR | Clear | | FPM | Feet Per Minute |
| CMD | Command | | FREQ | Frequency |
| CNI | Communication Navigation Identification | | FRZ | Freeze |
| COMM | Communications | | FTFP | Flight Test Function Panel |
| CONT | Contrast | | FW | Fire Warning Controls |
| | Control | | FX | Fixed |
| CRS | Course | | | |
| CTR | Center | | | **G** |
| CTXFR | Center Tank Transfer | | | |
| | | | G | Groundspeed |
| | **D** | | | Guard |
| | | | | Load Factor |
| D | Dummy | | GC | Gyrocompass |
| dc | Direct Current | | GEN | Generator |
| DECR | Decreases | | GR | Gear |
| DGRD | Degraded | | GREC | Guard Receiver |
| DIR | Direction | | GT | Ground Track |
| DME | Distance Measuring Equipment | | | |
| DN | Down | | | **H** |
| DP | Diphase | | | |
| DSPL | Display | | HDG | Heading |
| | | | HI | High |
| | **E** | | HP | High Pressure |
| | | | HPC | High Pressure Compressor |
| E | East | | HQ | Have Quick |
| | Emergency | | HSI | Horizontal Situation Indicator |
| EADI | Electronic Attitude Director Indicator | | HUD | Head Up Display |
| ECC | Engine Control Computer | | HYD | Hydraulic |
| ECS | Environmental Control System | | HZ | Hydrazine |
| ECU | Engine Control Units | | | |
| EDT | | | | **I** |
| EFAIL | Simulated Failure | | | |
| EGT | Exhaust Gas Temperature | | I/P | Identification of Position |
| EHSI | Electronic Horizontal Situation Indicator | | IB | Inboard |
| ELEC | Electrical | | IBIT | Initiated Built-In Test |
| EMER or | | | ICS | Intercommunication System |
| EMERG | Emergency | | IDENT | Identification |
| ENG | Engine | | IFF | Identify Friend or Foe |
| EPS | Emergency Power System | | INC | Increases |
| EPU | Emergency Power Units | | | Increment |
| EST | Estimated | | INOP | Inoperative |
| ETA | Estimated Time of Arrival | | INS | Inertial Navigation System |
| ETE | Estimated Time Enroute | | INST | Instrument |
| EXT | External | | INT | Interior |
| | | | INTL | International |
| | **F** | | INTMD | Intermediate |
| | | | INV | Invalid |
| F | Feed | | ISM | Increased Stall Margin |
| | Fuel | | | |
| FADEC | Full Authority Digital Electronic Control | | | **K** |
| FCS | Flight Control System | | | |
| FECU | Flutter Exciter Control Unit | | KCAS | Knots Calibrated Airspeed |
| FF | Fuel Flow | | kva | Kilo-volt Amp |
| FLT | Flight | | | |
| FMC | Fuel Management Computer | | | |

## L

| | |
|---|---|
| L | Left |
| LAT | Latitude |
| LBS | Pounds |
| LC | Left Center |
| LCD | Liquid Crystal Displays |
| LD | Load |
| LDG | Landing |
| LDIV | Left Fuel Diverter |
| LEF | Leading Edge Flaps |
| LF | Left Feed Tank |
| LG | Landing Gear |
| LGT | Light(s) |
| LH | Left Hand |
| LIM | Limit |
| LNCHR | Launcher |
| LO | Low |
| LONG | Longitude |
| LOS | Line Of Sight |
| LP | Low Pressure |
| LPT | Low Pressure Turbine |
| LQD | Liquid |
| LT | Light |
| LW | Left Wing |
| LWP | Left Wing Transfer Pump |
| LWR | Lower |

## M

| | |
|---|---|
| M | Mach |
| | Manual |
| | Menu |
| MAC | Mean Aerodynamic Chord |
| MALF | Malfunction |
| MAN | Manual |
| MAX | Maximum |
| MC | Master Caution |
| MEA | Minimum Enroute Altitude |
| MECH | Mechanical |
| MED | Medium |
| MHz | Mega-Hertz |
| MIC | Microphone |
| MIL | Military |
| MIN | Minimum |
| MISC | Miscellaneous |
| MLG | Main Landing Gear |
| MN | Mach Number |
| MPCD | Multipurpose Color Displays |
| MPD | Multipurpose Display |
| MPDP | Multipurpose Display Processor |
| MRM | Medium Range Missile |
| MSL | Mean Sea Level |
| MSN | Mission |

## N

| | |
|---|---|
| N | North |
| N1 | Fan speed |
| N2 | Compressor Speed |
| NAT | National |
| NAV | Navigation |
| NB | Narrow Band |
| NLG | Nose Landing Gear |
| NM | Nautical Miles |
| NMI | Nautical Miles Indicated |
| No. | Number |
| NORM | Normal |
| NOZ | Nozzle |
| NTM | Northrop Technical Manual |
| NWS | Nose Wheel Steering |

## O

| | |
|---|---|
| O | Open |
| O/S | Off Set |
| OAT | Outside Air Temperature |
| OB | Outboard |
| OBOGS | On-Board Oxygen Generating System |
| OFLY | Overfly |
| OPR | Operate |
| ORIDE | Override |
| OT | Outside Temperature |
| OVHT | Overheat |
| OVRD | Override |
| OXY | Oxygen |

## P

| | |
|---|---|
| P | Plain |
| PAV | Prototype Air Vehicle |
| PC | Primary Control Hydraulic System |
| PC-1 | Power Control System One |
| PC-2 | Power Control System Two |
| PLT | Pilot |
| POS | Position |
| PP | Present Position |
| PPH | Pounds Per Hour |
| PRESS | Pressure |
| PROG | Program |
| psi | Pounds per Square Inch |
| PSIA | Pounds per Square Inch - Absolute |
| PT | Pitch Trim |
| PTO | Power Take-off |
| PWR | Power |

## Q

| | |
|---|---|
| QNH | Altimeter Setting |

## R

| | |
|---|---|
| R | Right |
| R1 | Radio 1 |
| R2 | Radio 2 |
| RAD | Radio |
| RC | Right Center |
| RCD | Record |
| RDIS | Redistribution |
| RDIV | Right Fuel Diverter |
| RDY | Ready |
| REC | Receive |
| RECT | Rectifier |
| REJ | Reject |
| REQ | Required |
| RET | Reticle |
| RF | Right Feed Tank |
| RH | Right Hand |
| RLG | Ring Laser Gyro |
| RLS | Reservoir Level Sensing |
| RNG | Range |
| RPM | Revolutions Per Minute |
| RSET | Reset |
| RV | Receive Variable |
| RW | Right Wing |
| RWP | Right Wing Transfer Pump |

## S

| | |
|---|---|
| S | South |
| SEL | Select |
| SEN | Sensor |
| SH | Stored Heading |
| SHF | Shift |
| SIL | Silence |
| SMS | Stores Management System |
| SP | Sequence Point |
| SPS | Secondary Power System |
| SRM | Short Range Missile |
| SRVS | Services |
| STBY | Standby |
| STR | Selected Route Point |
| STR | Steer |
| SYM | Symbology |

## T

| | |
|---|---|
| T | True Airspeed |
| T-R | Transmit/Receive |
| T/M | Telemetry |
| TACAN | Tactical Air Navigation |
| TCN | TACAN |
| TD | Time Delay |
| TEF | Trailing Edge Flaps |
| TEMP | Temperature |
| TGT | Target |
| TITL | Titling |
| TM or T/M | Telemetry |
| TOA | Time of Arrival (?) |
| TOD | Time-Of-Day |
| TOT | Take Off Trim |
| | **Time on Target** (?) |
| TSD | Tactical Situation Display |
| TT2 | Inlet Total Temperature Sensor |
| TX | Transmit |

## U

| | |
|---|---|
| U1 | UHF 1 |
| U2 | UHF 2 |
| UFC | Upfront Control |
| UHF | Ultra High Frequency |
| UPR | Upper |
| UTIL | Utility |
| UTIL-1 | Left Utility Hydraulic Pump |
| UTIL-2 | Right Utility Hydraulic Pump |
| UTIL-A | Utility A |
| UTIL-B | Utility B |
| UTL | Utility Hydraulic Pressure |
| UTM | Universal Transverse Mercator |

## V

| | |
|---|---|
| V/W | Voice Warning |
| VGS | VMS Gain Select |
| VMC | Vehicle Management Computer |
| VMES | Vehicle Management Electronics Set |
| VMS | Vehicle Management System |
| VOL | Volume |
| VTR | Video Tape Recording |
| VV | Vertical Velocity |
| VVI | Vertical Velocity Indication |

## W

| | |
|---|---|
| W | West |
| | Wing |
| WARN | Warning |
| WB | Wide Band |
| WCA | Warning, Caution, and Advisory |
| WG | Wing |
| WGS | World Geodetic System |
| WLP | Warning Light Panel |
| WNG | Wing |
| WOD | Word-Of-Day |
| WPN | Weapon |
| WT | Weight |

## X

| | |
|---|---|
| X | X band |
| XFMR | Transformer |
| XFP1 | Transfer Pump Number 1 |
| XFP2 | Transfer Pump Number 2 |

## Y

| | |
|---|---|
| Y | Y Band |

## Z

| | |
|---|---|
| Z | Zero |

INDEX

A28-INDEX

Page numbers for illustrations in the alphabetical index are underlined.

ard To:
Manager ATF Logistics Programs
Orgn 8810 - Mail Zone 89
Northrop Corporation
Aircraft Division
One Northrop Avenue
Hawthorne California, 90250-3277

## YF-23A
## PUBLICATION
## CHANGE REQUEST

DATE _____

REQUEST NO. _____

REQUESTED BY _____

PUBLICATION NO. _____  PUB. TITLE _____

PUB. DATE _____  PUB. REV. DATE _____

APPLICABLE MODEL _____  APPLICABLE SERIAL NO. _____

SECT/WP/SWP   PAGE           PARA.        FIG.          TABLE                        ITEM
NO. _____  NO. _____  NO. _____  NO. _____  NO. _____  STEP/S _____  NO. _____

SYSTEM _____  COMPONENT NAME _____

COMPONENT                                COMPONENT PART
MANUFACTURER _____  OR DESIGNATION NO. _____

REQUESTOR'S ORGANIZATION/DEPT _____  PHONE _____

DESCRIPTION OF PUBLICATION CHANGE REQUEST:

RECOMMENDATIONS·

DISPOSITION.

SHEET _____ OF _____ SHEET S